Fundamentals of Biochemistry

Adrian Dean

R CALLISTO REFERENCE

www.callistoreference.com

Callisto Reference,
118-35 Queens Blvd., Suite 400,
Forest Hills, NY 11375, USA

Visit us on the World Wide Web at:
www.callistoreference.com

ISBN: 978-1-64116-541-9 (Hardback)

Cataloging-in-Publication Data

Fundamentals of biochemistry / Adrian Dean.
    p. cm.
Includes bibliographical references and index.
ISBN 978-1-64116-541-9
1. Biochemistry. 2. Biology. 3. Chemistry. 4. Medical sciences. I. Dean, Adrian.
QH345 .F86 2022
572--dc23

Table of Contents

Preface		**VII**
Chapter 1	**Understanding Biochemistry**	**1**
	• Biomolecules	7
Chapter 2	**Carbohydrates and their Metabolism**	**9**
	• Monosaccharide	18
	• Disaccharide	31
	• Polysaccharide	39
	• Carbohydrate Metabolism	54
Chapter 3	**Lipid Metabolism**	**60**
	• Lipid	60
	• Fatty Acids	69
	• Lipoproteins	85
	• Lipid Metabolism	93
	• Cholesterol	96
	• Triacylglycerol Metabolism	107
Chapter 4	**Protein Metabolism**	**121**
	• Protein	121
	• Protein Metabolism	158
	• Protein Synthesis	160
	• Protein Targeting	164
	• Protein Glycosylation	169
Chapter 5	**Nucleic Acid**	**171**
	• DNA	172
	• Mitochondrial DNA	178
	• RNA	179
Chapter 6	**Enzymes: A Comprehensive Study**	**190**
	• Working Principle of Enzymes	217
	• Enzyme Kinetics	218
	• Enzyme Inhibition	222
Permissions		
Index		

Preface

The study of chemical processes related to living organisms is referred to as biochemistry. It is a subfield of both biology and chemistry. It is divided into fields like molecular genetics, protein science, and metabolism. Biochemistry focuses on how biological molecules give rise to processes that occur within and between the living cells. It also helps in the study and understanding of tissues and organs of an organism. Biochemistry focuses on the structures, functions, and interactions of biological macromolecules such ad nucleic acids, carbohydrates, protein, and lipids. The topics included in this textbook on biochemistry are of utmost significance and bound to provide incredible insights to readers. Some of the diverse topics covered in this book address the varied branches that fall under this category. In this textbook, constant effort has been made to make the understanding of the difficult concepts of biochemistry as easy and informative as possible, for the readers.

To facilitate a deeper understanding of the contents of this book a short introduction of every chapter is written below:

Chapter 1- Biochemistry is a subset of biology and chemistry that deals with the study of chemical processes within the living organisms. A biomolecule refers to the molecules and ions found in an organism. These biomolecules play a crucial role in facilitating biological processes such as morphogenesis, cell division, etc. This is an introductory chapter which will introduce briefly all the significant aspects of biochemistry.

Chapter 2- A carbohydrate is a biomolecule that comprises of atoms of hydrogen, carbon and oxygen. It has four groupings known as disaccharides, monosaccharides, polysaccharides and oligosaccharides. The various biochemical processes responsible for the metabolic formation, breakdown and interconversion of carbohydrates in living organisms is referred to as carbohydrate metabolism. This chapter has been carefully written to provide an easy understanding of the varied facets of carbohydrates and its metabolism.

Chapter 3- A biomolecule that is soluble in nonpolar solvents is known as lipid. The various categories of lipids include glycerophospholipids, fatty acids, sterol lipids, polyketides, prenol lipids, glycerolipids, sphingolipids and saccharolipids. The synthesis and degradation of lipids in cell is known as lipid metabolism. The chapter closely examines the diverse aspects of lipids and the key concepts of lipid metabolism to provide an extensive understanding of the subject.

Chapter 4- Proteins are the macromolecules that comprise of one or more than one long chain of amino acid residues. It performs functions like DNA replication, responding to stimuli, catalysing metabolic reactions, provide structure to organisms and cells, transport molecules, etc. The various biomolecule processes responsible for the breakdown of proteins and the synthesis of proteins and amino acids is referred to as protein metabolism. The topics elaborated in this chapter will help in gaining a better perspective about the functions of protein and their protein metabolism.

Chapter 5- The overall name for DNA and RNA is known as nucleic acids. They are the small biomolecules that are composed of nucleotides which are made up of three components: 5-a carbon sugar, nitrogenous base and a phosphate group. The topics elaborated in this chapter will help in gaining a better perspective about the nucleic acids.

Chapter 6- The macromolecular biological catalysts that accelerate chemical reactions are known as enzymes. Substrates are the molecules upon which enzymes act and convert it into another molecules known as products. The studies of the chemical reactions that are catalysed by enzymes fall under the domain of enzyme kinetics. This chapter has been carefully written to provide an easy understanding of the various types of enzymes.

I would like to share the credit of this book with my editorial team who worked tirelessly on this book. I owe the completion of this book to the never-ending support of my family, who supported me throughout the project.

Adrian Dean

Chapter 1

Understanding Biochemistry

Biochemistry is a subset of biology and chemistry that deals with the study of chemical processes within the living organisms. A biomolecule refers to the molecules and ions found in an organism. These biomolecules play a crucial role in facilitating biological processes such as morphogenesis, cell division, etc. This is an introductory chapter which will introduce briefly all the significant aspects of biochemistry.

Biochemistry is the study of the chemical substances and processes that occur in plants, animals, and microorganisms and of the changes they undergo during development and life. It deals with the chemistry of life, and as such it draws on the techniques of analytical, organic, and physical chemistry, as well as those of physiologists concerned with the molecular basis of vital processes. All chemical changes within the organism—either the degradation of substances, generally to gain necessary energy, or the buildup of complex molecules necessary for life processes—are collectively termed metabolism. These chemical changes depend on the action of organic catalysts known as enzymes, and enzymes, in turn, depend for their existence on the genetic apparatus of the cell. It is not surprising, therefore, that biochemistry enters into the investigation of chemical changes in disease, drug action, and other aspects of medicine, as well as in nutrition, genetics, and agriculture.

The term biochemistry is synonymous with two somewhat older terms: physiological chemistry and biological chemistry. Those aspects of biochemistry that deal with the chemistry and function of very large molecules (e.g., proteins and nucleic acids) are often grouped under the term molecular biology.

Areas of Study

A description of life at the molecular level includes a description of all the complexly interrelated chemical changes that occur within the cell—i.e., the processes known as intermediary metabolism. The processes of growth, reproduction, and heredity, also subjects of the biochemist's curiosity, are intimately related to intermediary metabolism and cannot be understood independently of it. The properties and capacities exhibited by a complex multicellular organism can be reduced to the properties of the individual cells of that organism, and the behaviour of each individual cell can be understood in terms of its chemical structure and the chemical changes occurring within that cell.

Chemical Composition of Living Matter

Every living cell contains, in addition to water and salts or minerals, a large number of organic compounds, substances composed of carbon combined with varying amounts of hydrogen and usually also of oxygen. Nitrogen, phosphorus, and sulfur are likewise common constituents. In general, the bulk of the organic matter of a cell may be classified as (1) protein, (2) carbohydrate, and (3) fat, (4) lipid. Nucleic acids and various other organic derivatives are also important constituents. Each class contains a great diversity of individual compounds. Many substances that cannot be classified in any of the above categories also occur, though usually not in large amounts.

Proteins are fundamental to life, not only as structural elements (e.g., collagen) and to provide defense (as antibodies) against invading destructive forces but also because the essential biocatalysts are proteins.

Fats, or lipids, constitute a heterogeneous group of organic chemicals that can be extracted from biological material by nonpolar solvents such as ethanol, ether, and benzene. The liver is the main site of fat metabolism.The control of fat absorption is known to depend upon a combination action of secretions of the pancreas and bile salts. Abnormalities of fat metabolism, which result in disorders such as obesity and rare clinical conditions, are the subject of much biochemical research. Equally interesting to biochemists is the association between high levels of fat in the blood and the occurrence of arteriosclerosis ("hardening" of the arteries).

Nucleic acids are large, complex compounds of very high molecular weight present in the cells of all organisms and in viruses. They are of great importance in the synthesis of proteins and in the transmission of hereditary information from one generation to the next. Originally discovered as constituents of cell nuclei, it was assumed for many years after their isolation in 1869 that they were found nowhere else. This assumption was not challenged seriously until the 1940s, when it was determined that two kinds of nucleic acid exist: deoxyribonucleic acid (DNA), in the nuclei of all cells and in some viruses; and ribonucleic acid (RNA), in the cytoplasm of all cells and in most viruses.

Nutrition

Biochemists have long been interested in the chemical composition of the food of animals. All animals require organic material in their diet, in addition to water and minerals. This organic matter must be sufficient in quantity to satisfy the caloric, or energy, requirements of the animals. Within certain limits, carbohydrate, fat, and protein may be used interchangeably for this purpose. In addition, however, animals have nutritional requirements for specific organic compounds. Certain essential fatty acids, about ten different amino acids (the so-called essential amino acids), and vitamins are required by many higher animals. The nutritional requirements of various species are similar but not necessarily identical; thus man and the guinea pig require vitamin C, or ascorbic acid, whereas the rat does not.

That plants differ from animals in requiring no preformed organic material was appreciated soon after the plant studies of the late 1700s. The ability of green plants to make all their cellular material from simple substances—carbon dioxide, water, salts, and a source of nitrogen such as ammonia or nitrate—was termed photosynthesis. As the name implies, light is required as an energy source, and it is generally furnished by sunlight. The process itself is primarily concerned with the manufacture of carbohydrate, from which fat can be made by animals that eat plant carbohydrates. Protein can also be formed from carbohydrate, provided ammonia is furnished.

In spite of the large apparent differences in nutritional requirements of plants and animals, the patterns of chemical change within the cell are the same. The plant manufactures all the materials it needs, but these materials are essentially similar to those that the animal cell uses and are often handled in the same way once they are formed. Plants could not furnish animals with their nutritional requirements if the cellular constituents in the two forms were not basically similar.

Digestion

The organic food of animals, including man, consists in part of large molecules. In the digestive tracts of higher animals, these molecules are hydrolyzed, or broken down, to their component building blocks. Proteins are converted to mixtures of amino acids, and polysaccharides are converted to monosaccharides. In general, all living forms use the same small molecules, but many of the large complex molecules are different in each species. An animal, therefore, cannot use the protein of a plant or of another animal directly but must first break it down to amino acids and then recombine the amino acids into its own characteristic proteins. The hydrolysis of food material is necessary also to convert solid material into soluble substances suitable for absorption. The liquefaction of stomach contents aroused the early interest of observers, long before the birth of modern chemistry, and the hydrolytic enzymes secreted into the digestive tract were among the first enzymes to be studied in detail. Pepsin and trypsin, the proteolytic enzymes of gastric and pancreatic juice, respectively, continue to be intensively investigated.

The products of enzymatic action on the food of an animal are absorbed through the walls of the intestines and distributed to the body by blood and lymph. In organisms without digestive tracts, substances must also be absorbed in some way from the environment. In some instances simple diffusion appears to be sufficient to explain the transfer of a substance across a cell membrane. In other cases, however (e.g., in the case of the transfer of glucose from the lumen of the intestine to the blood), transfer occurs against a concentration gradient. That is, the glucose may move from a place of lower concentration to a place of higher concentration.

In the case of the secretion of hydrochloric acid into gastric juice, it has been shown that active secretion is dependent on an adequate oxygen supply (i.e., on the respiratory metabolism of the tissue), and the same holds for absorption of salts by plant roots. The energy released during the tissue oxidation must be harnessed in some way to provide the energy necessary for the absorption or secretion. This harnessing is achieved by a special chemical coupling system. The elucidation of the nature of such coupling systems has been an objective of the biochemist.

Blood

One of the animal tissues that has always excited special curiosity is blood.

The blood pigment hemoglobin has been intensively studied. Hemoglobin is confined within the blood corpuscles and carries oxygen from the lungs to the tissues. It combines with oxygen in the lungs, where the oxygen concentration is high, and releases the oxygen in the tissues, where the oxygen concentration is low. The hemoglobins of higher animals are related but not identical. In invertebrates, other pigments may take the place and function of hemoglobin.

The proteins of blood plasma also have been extensively investigated. The gamma-globulin fraction of the plasma proteins contains the antibodies of the blood and is of practical value as an immunizing agent. An animal develops resistance to disease largely by antibody production. Antibodies are proteins with the ability to combine with an antigen (i.e., an agent that induces their formation). When this agent is a component of a disease-causing bacterium, the antibody can protect an organism from infection by that bacterium. The chemical study of antigens and antibodies and their interrelationship is known as immunochemistry.

Metabolism and Hormones

The cell is the site of a constant, complex, and orderly set of chemical changes collectively called metabolism. Metabolism is associated with a release of heat. The heat released is the same as that obtained if the same chemical change is brought about outside the living organism. This confirms the fact that the laws of thermodynamics apply to living systems just as they apply to the inanimate world. The pattern of chemical change in a living cell, however, is distinctive and different from anything encountered in nonliving systems. This difference does not mean that any chemical laws are invalidated. It instead reflects the extraordinary complexity of the interrelations of cellular reactions.

Hormones, which may be regarded as regulators of metabolism, are investigated at three levels, to determine (1) their physiological effects, (2) their chemical structure, and (3) the chemical mechanisms whereby they operate. The study of the physiological effects of hormones is properly regarded as the province of the physiologist. Such investigations obviously had to precede the more analytical chemical studies. The chemical structures of thyroxine and adrenaline are known. The chemistry of the sex and adrenal hormones, which are steroids, has also been thoroughly investigated. The hormones of the pancreas—insulin and glucagon—and the hormones of the hypophysis (pituitary gland) are peptides (i.e., compounds composed of chains of amino acids). The structures of most of these hormones has been determined. The chemical structures of the plant hormones, auxin and gibberellic acid, which act as growth-controlling agents in plants, are also known.

The first and second phases of the hormone problem thus have been well, though not completely, explored, but the third phase is still in its infancy. It seems likely that different hormones exert their effects in different ways. Some may act by affecting the permeability of membranes; others appear to control the synthesis of certain enzymes. Evidently some hormones also control the activity of certain genes.

Genes

Genetic studies have shown that the hereditary characteristics of a species are maintained and transmitted by the self-duplicating units known as genes, which are composed of nucleic acids and located in the chromosomes of the nucleus., the capacity of a protein to behave as an enzyme is determined by the chemical constitution of the gene (DNA) that directs the synthesis of the protein. The relationship of genes to enzymes has been demonstrated in several ways.

Evolution and Origin of Life

The exploration of space beginning in the mid-20th century intensified speculation about the possibility of life on other planets. At the same time, man was beginning to understand some of the intimate chemical mechanisms used for the transmission of hereditary characteristics. It was possible, by studying protein structure in different species, to see how the amino acid sequences of functional proteins (e.g., hemoglobin and cytochrome) have been altered during phylogeny (the development of species). It was natural, therefore, that biochemists should look upon the problem of the origin of life as a practical one. The synthesis of a living cell from inanimate material was not regarded as an impossible task for the future.

Methods in Biochemistry

Like other sciences, biochemistry aims at quantifying, or measuring, results, sometimes with sophisticated instrumentation. The earliest approach to a study of the events in a living organism was an analysis of the materials entering an organism (foods, oxygen) and those leaving (excretion products, carbon dioxide). This is still the basis of so-called balance experiments conducted on animals, in which, for example, both foods and excreta are thoroughly analyzed. For this purpose many chemical methods involving specific colour reactions have been developed, requiring spectrum-analyzing instruments (spectrophotometers) for quantitative measurement. Gasometric techniques are those commonly used for measurements of oxygen and carbon dioxide, yielding respiratory quotients (the ratio of carbon dioxide to oxygen).Because these techniques yield an overall picture of metabolic capacities, it became necessary to disrupt cellular structure (homogenization) and to isolate the individual parts of the cell—nuclei, mitochondria, lysosomes, ribosomes, membranes—and finally the various enzymes and discrete chemical substances of the cell in an attempt to understand the chemistry of life more fully.

Centrifugation and Electrophoresis

An important tool in biochemical research is the centrifuge, which through rapid spinning imposes high centrifugal forces on suspended particles, or even molecules in solution, and causes separations of such matter on the basis of differences in weight. Thus, red cells may be separated from plasma of blood, nuclei from mitochondria in cell homogenates, and one protein from another in complex mixtures. Proteins are separated by ultracentrifugation—very high speed spinning; with appropriate photography of the protein layers as they form in the centrifugal field, it is possible to determine the molecular weights of proteins.

Another property of biological molecules that has been exploited for separation and analysis is their electrical charge. Amino acids and proteins possess net positive or negative charges according to the acidity of the solution in which they are dissolved. In an electric field, such molecules adopt different rates of migration toward positively (anode) or negatively (cathode) charged poles and permit separation. Such separations can be effected in solutions or when the proteins saturate a stationary medium such as cellulose (filter paper), starch, or acrylamide gels. By appropriate colour reactions of the proteins and scanning of colour intensities, a number of proteins in a mixture may be measured. Separate proteins may be isolated and identified by electrophoresis, and the purity of a given protein may be determined. (Electrophoresis of human hemoglobin revealed the abnormal hemoglobin in sickle-cell anemia, the first definitive example of a "molecular disease.")

Chromatography and Isotopes

The different solubilities of substances in aqueous and organic solvents provide another basis for analysis. In its earlier form, a separation was conducted in complex apparatus by partition of substances in various solvents. A simplified form of the same principle evolved as "paper chromatography," in which small amounts of substances could be separated on filter paper and identified by appropriate colour reactions. In contrast to electrophoresis, this method has been applied to a wide variety of biological compounds and has contributed enormously to research in biochemistry.

The general principle has been extended from filter paper strips to columns of other relatively inert media, permitting larger scale separation and identification of closely related biological substances. Another technique of column chromatography is based on the relative rates of penetration of molecules into beads of a complex carbohydrate according to size of the molecules. Larger molecules are excluded relative to smaller molecules and emerge first from a column of such beads. This technique not only permits separation of biological substances but also provides estimates of molecular weights.

Perhaps the single most important technique in unravelling the complexities of metabolism has been the use of isotopes (heavy or radioactive elements) in labelling biological compounds and "tracing" their fate in metabolism. Measurement of the isotope-labelled compounds has required considerable technology in mass spectroscopy and radioactive detection devices.

A variety of other physical techniques, such as nuclear magnetic resonance, electron spin spectroscopy, circular dichroism, and X-ray crystallography, have become prominent tools in revealing the relation of chemical structure to biological function.

Plant Biochemistry

Plant biochemistry is the study of the biochemistry of autotrophic organisms such as photosynthesis and other plant specific biochemical processes.

The plants and higher fungi produce through their metabolism a vast variety of chemical substances. These are important for the plant itself, but also for the environment and for the recovery and use by humans. The Plant Biochemistry deals with biochemical processes of plant metabolism. The entirety of the vital processes of plants is also known as plant physiology.

The plant biochemistry is therefore a branch of Biochemistry.

Phytochemistry

Phyrochemistry is the study of phytochemicals, these are secondary metabolic substances found in plants. Many of these are known to provide protection against insect attacks and plant diseases. They also exhibit a number of protective functions for human consumers.

Phytochemistry is widely used in the field of herbal medicine.

Phytochemical technique mainly applies to the quality control of Chinese medicine or herbal medicine of various chemical components, such as saponins, alkaloids, volatile oils, flavonoids and anthraquinones. In most cases, biologically active compounds in herbal medicine have not been determined. Therefore, it is important to use the phytochemical methods to screen and analyze bioactive components, not only for the quality control of crudedrugs, but also for the elucidation of their therapeutic mechanisms.

Animal Biochemistry

Animal biochemistry is a branch of science that studies about structure and function cellular components such as protein, carbohydrate, lipid, nucleic acids, and other bio-molecules in animals. In veterinary education and research, biochemistry is highly relevant to the metabolism and functions of animals in health and disease, and forms the basis for an intelligent understanding of major aspects of veterinary science and animal husbandry.

Biomolecules

Biomolecules are defined as any organic molecule present in a living cell which includes carbohydrates, proteins, fats etc. Each biomolecule is essential for body functions and manufactured within the body. They can vary in nature, type, and structure where some may be straight chains, some may be cyclic rings or both. Also, they can vary in physical properties such as water solubility, melting points.

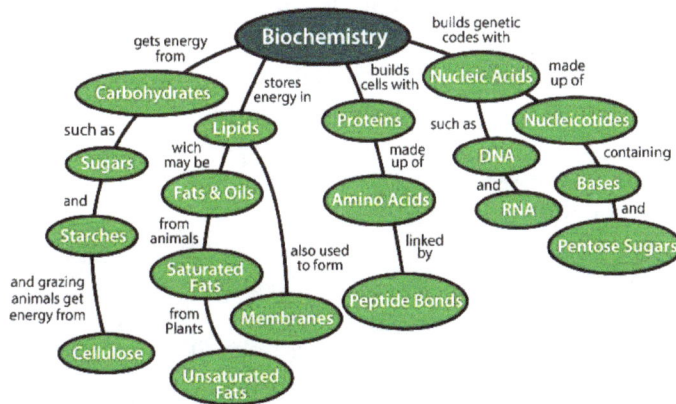

Types of Biomolecules

Biomolecules are primarily classified into 4 types, namely:

Carbohydrates

Polysaccharides, commonly known as carbohydrates are macromolecules. They are made up of monosaccharides (sugar molecules). Majority of living cells are rich in carbohydrates and they are the final products of many metabolisms. For example, Glucose is the final product of photosynthesis. Saccharides can be monosaccharide, disaccharide, polysaccharide etc. based on the number of sugar molecules they are made up of.

Proteins

Proteins are dietary compounds made of monomers called amino acids. Protein is a long chain of amino acid bonded by polypeptide bonds. Hence proteins are also called polypeptides. Amino acids are carbon-containing compounds where a carboxylic acid group and the amino group are present at the two ends. Each amino acid consists of one central carbon surrounded by four substituents. These four substituents include an amino group, carboxylic acid group, hydrogen and a variable group represented by R. The variable group, R decides the nature and type of amino acid.

Lipids

Lipids are a group of water-insoluble compounds which include fats, glycerol, phospholipids, steroids, oils etc. Types of lipids vary according to their constituents. Fatty acids are simple lipids made up of a carboxyl group and a variable group R. They may be saturated or unsaturated fatty acids. Glycerol is trihydroxy propane which combines with fatty acids to give triglycerides. Some lipids consist of a phosphorus group along with the organic chain. Such lipids are called phospholipids, the constituent of the plasma membrane.

Nucleic Acids

Nucleic acids are the genetic materials present in an organism which includes DNA and RNA. Nucleic acids are the combination materials of nitrogenous bases, sugar molecules and phosphate group linked by different bonds in a series of steps. Our body consists of heterocyclic compounds like pyrimidines and purines. These are nitrogenous compounds like adenine, guanine, cytosine, thymine, and uracil. When these bases bond with sugar chains, nucleosides are formed. Nucleosides in turn bond with a phosphate group to give nucleotides like DNA and RNA.

The human body consists of trillions of cells which are made up of carbohydrates, proteins like biomolecules. Majority of cell activities depend on them.

References

- Biochemistry, science: britannica.com, Retrieved January 4, 2019

- Plant-biochemistry, chemistry : internetchemistry.com, Retrieved February 15, 2019

- What-is-Phytochemistry, What-is-Phytochemistry, Phyrochemistry: scribd.com, Retrieved January 9, 2019

- Animals-biochemistry, events-list: alliedacademies.com, Retrieved February 27, 2019

- Biomolecules-in-living-organisms, biology: byjus.com, Retrieved May 3, 2019

Chapter 2

Carbohydrates and their Metabolism

A carbohydrate is a biomolecule that comprises of atoms of hydrogen, carbon and oxygen. It has four groupings known as disaccharides, monosaccharides, polysaccharides and oligosaccharides. The various biochemical processes responsible for the metabolic formation, breakdown and inter-conversion of carbohydrates in living organisms is referred to as carbohydrate metabolism. This chapter has been carefully written to provide an easy understanding of the varied facets of carbohydrates and its metabolism.

Carbohydrate is the class of naturally occurring compounds and derivatives formed from them The general formula $C_x(H_2O)_y$ is commonly used to represent many carbohydrates, which means "watered carbon."

Carbohydrates are probably the most abundant and widespread organic substances in nature, and they are essential constituents of all living things. Carbohydrates are formed by green plants from carbon dioxide and water during the process of photosynthesis. Carbohydrates serve as energy sources and as essential structural components in organisms; in addition, part of the structure of nucleic acids, which contain genetic information, consists of carbohydrate.

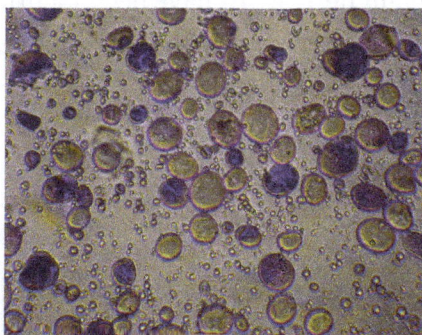

Wheat starch granules stained with iodine.

General Features

Classification and Nomenclature

Although a number of classification schemes have been devised for carbohydrates, the division into four major groups—monosaccharides, disaccharides, oligosaccharides, and polysaccharides—used here is among the most common. Most monosaccharides, or simple sugars, are found in grapes, other fruits, and honey. Although they can contain from three to nine carbon atoms, the most common representatives consist of five or six joined together to form a chainlike molecule. Three of the most important simple sugars—glucose (also known as dextrose, grape sugar, and corn sugar), fructose (fruit sugar), and galactose—have the same molecular formula, $(C_6H_{12}O_6)$, but, because their atoms have different structural arrangements, the sugars have different characteristics; i.e., they are isomers.

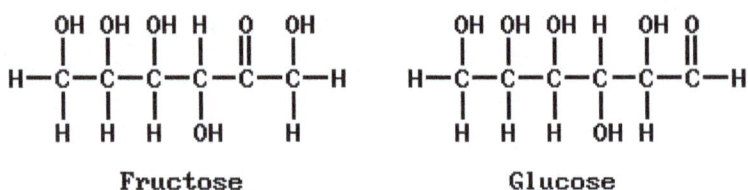

Fructose   Glucose

Slight changes in structural arrangements are detectable by living things and influence the biological significance of isomeric compounds. It is known, for example, that the degree of sweetness of various sugars differs according to the arrangement of the hydroxyl groups ($-OH$) that compose part of the molecular structure. A direct correlation that may exist between taste and any specific structural arrangement, however, has not yet been established; that is, it is not yet possible to predict the taste of a sugar by knowing its specific structural arrangement. The energy in the chemical bonds of glucose indirectly supplies most living things with a major part of the energy that is necessary for them to carry on their activities. Galactose, which is rarely found as a simple sugar, is usually combined with other simple sugars in order to form larger molecules.

Two molecules of a simple sugar that are linked to each other form a disaccharide, or double sugar. The disaccharide sucrose, or table sugar, consists of one molecule of glucose and one molecule of fructose; the most familiar sources of sucrose are sugar beets and cane sugar. Milk sugar, or lactose, and maltose are also disaccharides. Before the energy in disaccharides can be utilized by living things, the molecules must be broken down into their respective monosaccharides. Oligosaccharides, which consist of three to six monosaccharide units, are rather infrequently found in natural sources, although a few plant derivatives have been identified.

Lactose crystal

Lactose crystals are shown suspended in oil. Their distinct shape allows them to be identified in foods examined for research.

Polysaccharides (the term means many sugars) represent most of the structural and energy-reserve carbohydrates found in nature. Large molecules that may consist of as many as 10,000 monosaccharide units linked together, polysaccharides vary considerably in size, in structural complexity, and in sugar content; several hundred distinct types have thus far been identified. Cellulose, the principal structural component of plants, is a complex polysaccharide comprising many glucose units linked together; it is the most common polysaccharide. The starch found in plants and the glycogen found in animals also are complex glucose polysaccharides. Starch (from the Old English word stercan, meaning "to stiffen") is found mostly in seeds, roots, and stems, where it is stored as an available energy source for plants. Plant starch may be processed into foods such as bread, or it may be consumed directly—as in potatoes, for instance. Glycogen, which consists of branching chains of glucose molecules, is formed in the liver and muscles of higher animals and is stored as an energy source.

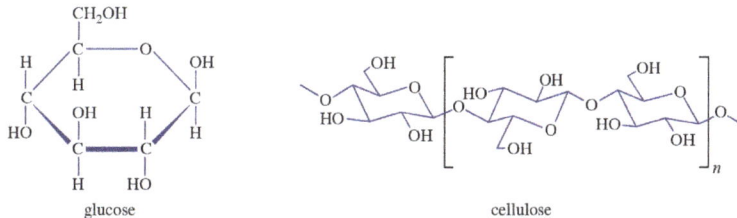

composition of cellulose and glucose

Cellulose and glucose are examples of carbohydrates.

The generic nomenclature ending for the monosaccharides is -ose; thus, the term pentose (pent = five) is used for monosaccharides containing five carbon atoms, and hexose (hex = six) is used for those containing six. In addition, because the monosaccharides contain a chemically reactive group that is either an aldehyde group or a keto group, they are frequently referred to as aldopentoses or ketopentoses or aldohexoses or ketohexoses. The aldehyde group can occur at position 1 of an aldopentose, and the keto group can occur at a further position within a ketohexose. Glucose is an aldohexose—i.e., it contains six carbon atoms, and the chemically reactive group is an aldehyde group.

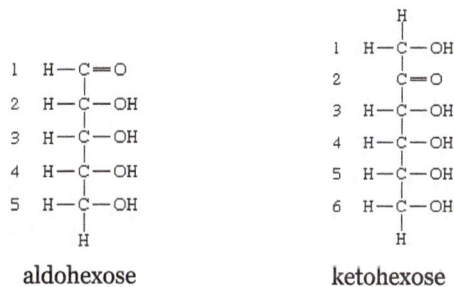

aldohexose   ketohexose

Biological Significance

The importance of carbohydrates to living things can hardly be overemphasized. The energy stores of most animals and plants are both carbohydrate and lipid in nature; carbohydrates are generally available as an immediate energy source, whereas lipids act as a long-term energy resource and tend to be utilized at a slower rate. Glucose, the prevalent uncombined, or free, sugar circulating in the blood of higher animals, is essential to cell function. The proper regulation of glucose metabolism is of paramount importance to survival.

The ability of ruminants, such as cattle, sheep, and goats, to convert the polysaccharides present in grass and similar feeds into protein provides a major source of protein for humans. A number of medically important antibiotics, such as streptomycin, are carbohydrate derivatives. The cellulose in plants is used to manufacture paper, wood for construction, and fabrics.

Role in the Biosphere

The essential process in the biosphere, the portion of Earth in which life can occur, that has permitted the evolution of life as it now exists is the conversion by green plants of carbon dioxide from the atmosphere into carbohydrates, using light energy from the Sun. This process, called photosynthesis, results in both the release of oxygen gas into the atmosphere and the transformation of light energy into the chemical energy of carbohydrates. The energy stored by plants during the formation of carbohydrates is used by animals to carry out mechanical work and to perform biosynthetic activities.

During photosynthesis, an immediate phosphorous-containing product known as 3-phosphoglyceric acid is formed.

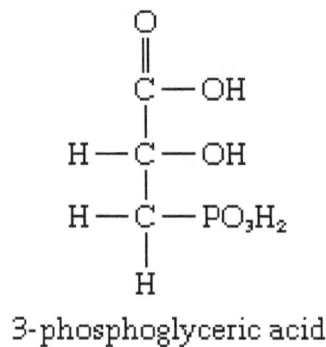

$$
\begin{array}{c}
\text{O} \\
\parallel \\
\text{C} - \text{OH} \\
| \\
\text{H} - \text{C} - \text{OH} \\
| \\
\text{H} - \text{C} - \text{PO}_3\text{H}_2 \\
| \\
\text{H}
\end{array}
$$

3-phosphoglyceric acid

This compound then is transformed into cell wall components such as cellulose, varying amounts of sucrose, and starch—depending on the plant type—and a wide variety of polysaccharides, other than cellulose and starch, that function as essential structural components.

Role in Human Nutrition

The total caloric, or energy, requirement for an individual depends on age, occupation, and other factors but generally ranges between 2,000 and 4,000 calories per 24-hour period (one calorie, as this term is used in nutrition, is the amount of heat necessary to raise the temperature of 1,000 grams of water from 15 to 16 °C (59 to 61 °F); in other contexts this amount of heat is called the kilocalorie). Carbohydrate that can be used by humans produces four calories per gram as opposed to nine calories per gram of fat and four per gram of protein. In areas of the world where nutrition is marginal, a high proportion (approximately one to two pounds) of an individual's daily energy requirement may be supplied by carbohydrate, with most of the remainder coming from a variety of fat sources.

Although carbohydrates may compose as much as 80 percent of the total caloric intake in the human diet, for a given diet, the proportion of starch to total carbohydrate is quite variable, depending upon the prevailing customs. In East Asia and in areas of Africa, for example, where rice or tubers such as manioc provide a major food source, starch may account for as much as 80 percent of the total carbohydrate intake. In a typical Western diet, 33 to 50 percent of the caloric intake is in the form of carbohydrate. Approximately half (i.e., 17 to 25 percent) is represented by starch; another third by table sugar (sucrose) and milk sugar (lactose); and smaller percentages by monosaccharides such as glucose and fructose, which are common in fruits, honey, syrups, and certain vegetables such as artichokes, onions, and sugar beets. The small remainder consists of bulk, or indigestible carbohydrate, which comprises primarily the cellulosic outer covering of seeds and the stalks and leaves of vegetables.

Role in Energy Storage

Starches, the major plant-energy-reserve polysaccharides used by humans, are stored in plants in the form of nearly spherical granules that vary in diameter from about three to 100 micrometres (about 0.0001 to 0.004 inch). Most plant starches consist of a mixture of two components: amylose and amylopectin. The glucose molecules composing amylose have a straight-chain, or linear, structure. Amylopectin has a branched-chain structure and is a somewhat more compact molecule. Several thousand glucose units may be present in a single starch molecule. (In the diagram, each small circle represents one glucose molecule).

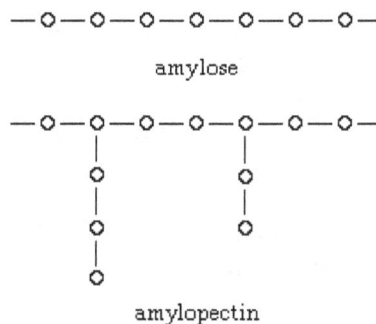

amylose

amylopectin

In addition to granules, many plants have large numbers of specialized cells, called parenchymatous cells, the principal function of which is the storage of starch; examples of plants with these cells include root vegetables and tubers. The starch content of plants varies considerably; the highest concentrations are found in seeds and in cereal grains, which contain up to 80 percent of their total carbohydrate as starch. The amylose and amylopectin components of starch occur in variable proportions; most plant species store approximately 25 percent of their starch as amylose and 75 percent as amylopectin. This proportion can be altered, however, by selective-breeding techniques, and some varieties of corn have been developed that produce up to 70 percent of their starch as amylose, which is more easily digested by humans than is amylopectin.

In addition to the starches, some plants (e.g., the Jerusalem artichoke and the leaves of certain grasses, particularly rye grass) form storage polysaccharides composed of fructose units rather than glucose. Although the fructose polysaccharides can be broken down and used to prepare syrups, they cannot be digested by higher animals.

Starches are not formed by animals; instead, they form a closely related polysaccharide, glycogen. Virtually all vertebrate and invertebrate animal cells, as well as those of numerous fungi and protozoans, contain some glycogen; particularly high concentrations of this substance are found in the liver and muscle cells of higher animals. The overall structure of glycogen, which is a highly branched molecule consisting of glucose units, has a superficial resemblance to that of the amylopectin component of starch, although the structural details of glycogen are significantly different.

Under conditions of stress or muscular activity in animals, glycogen is rapidly broken down to glucose, which is subsequently used as an energy source. In this manner, glycogen acts as an immediate carbohydrate reserve. Furthermore, the amount of glycogen present at any given time, especially in the liver, directly reflects an animal's nutritional state. When adequate food supplies are available, both glycogen and fat reserves of the body increase, but when food supplies decrease or when the food intake falls below the minimum energy requirements, the glycogen reserves are depleted quite rapidly, while those of fat are used at a slower rate.

Role in Plant and Animal Structure

Whereas starches and glycogen represent the major reserve polysaccharides of living things, most of the carbohydrate found in nature occurs as structural components in the cell walls of plants. Carbohydrates in plant cell walls generally consist of several distinct layers, one of which contains a higher concentration of cellulose than the others. The physical and chemical properties of cellulose are strikingly different from those of the amylose component of starch.

In most plants, the cell wall is about 0.5 micrometre thick and contains a mixture of cellulose, pentose-containing polysaccharides (pentosans), and an inert (chemically unreactive) plastic-like material called lignin. The amounts of cellulose and pentosan may vary; most plants contain between 40 and 60 percent cellulose, although higher amounts are present in the cotton fibre.

Polysaccharides also function as major structural components in animals. Chitin, which is similar to cellulose, is found in insects and other arthropods. Other complex polysaccharides predominate in the structural tissues of higher animals.

Structural Arrangements and Properties

Stereoisomerism

Studies by German chemist Emil Fischer in the late 19th century showed that carbohydrates, such as fructose and glucose, with the same molecular formulas but with different structural arrangements and properties (i.e., isomers) can be formed by relatively simple variations of their spatial, or geometric, arrangements. This type of isomerism, which is called stereoisomerism, exists in all biological systems. Among carbohydrates, the simplest example is provided by the three-carbon aldose sugar glyceraldehyde. There is no way by which the structures of the two isomers of glyceraldehyde, which can be distinguished by the so-called Fischer projection formulas, can be made identical, excluding breaking and reforming the linkages, or bonds, of the hydrogen ($-H$) and hydroxyl ($-OH$) groups attached to the carbon at position 2. The isomers are, in fact, mirror images akin to right and left hands; the term enantiomorphism is frequently employed for such isomerism. The chemical and physical properties of enantiomers are identical except for the property of optical rotation.

Optical rotation is the rotation of the plane of polarized light. Polarized light is light that has been separated into two beams that vibrate at right angles to each other; solutions of substances that rotate the plane of polarization are said to be optically active, and the degree of rotation is called the optical rotation of the solution. In the case of the isomers of glyceraldehyde, the magnitudes of the optical rotation are the same, but the direction in which the light is rotated—generally designated as plus, or d for dextrorotatory (to the right), or as minus, or l for levorotatory (to the left)—is opposite; i.e., a solution of D-(d)-glyceraldehyde causes the plane of polarized light to rotate to the right, and a solution of L-(l)-glyceraldehyde rotates the plane of polarized light to the left. Fischer projection formulas for the two isomers of glyceraldehyde are given below.

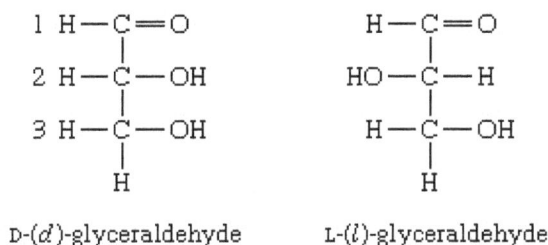

$$1 \; H-C=O \qquad\qquad H-C=O$$
$$2 \; H-C-OH \qquad HO-C-H$$
$$3 \; H-C-OH \qquad\quad H-C-OH$$
$$H \qquad\qquad\qquad H$$

D-(d)-glyceraldehyde   L-(l)-glyceraldehyde

Configuration

Molecules, such as the isomers of glyceraldehyde—the atoms of which can have different structural arrangements—are known as asymmetrical molecules. The number of possible structural arrangements for an asymmetrical molecule depends on the number of centres of asymmetry; i.e., for n (any given number of) centres of asymmetry, 2^n different isomers of a molecule are possible. An asymmetrical centre in the case of carbon is defined as a carbon atom to which four different groups are attached. In the three-carbon aldose sugar, glyceraldehyde, the asymmetrical centre is located at the central carbon atom.

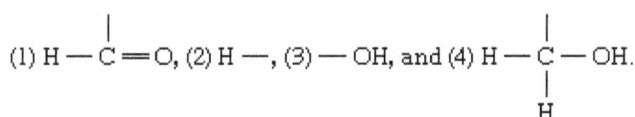

$$(1) \; H-C=O, (2) \; H-, (3) -OH, \text{ and } (4) \; H-C-OH.$$
$$H$$

The position of the hydroxyl group (−OH) attached to the central carbon atom—i.e., whether −OH projects from the left or the right—determines whether the molecule rotates the plane of polarized light to the left or to the right. Since glyceraldehyde has one asymmetrical centre, n is one in the relationship 2^n, and there thus are two possible glyceraldehyde isomers. Sugars containing four carbon atoms have two asymmetrical centres; hence, there are four possible isomers (2^2). Similarly, sugars with five carbon atoms have three asymmetrical centres and thus have eight possible isomers (2^3). Keto sugars have one less asymmetrical centre for a given number of carbon atoms than do aldehyde sugars.

A convention of nomenclature, devised in 1906, states that the form of glyceraldehyde whose asymmetrical carbon atom has a hydroxyl group projecting to the right is designated as of the D-configuration; that form, whose asymmetrical carbon atom has a hydroxyl group projecting to the left, is designated as L. All sugars that can be derived from D-glyceraldehyde—i.e., hydroxyl group attached to the asymmetrical carbon atom most remote from the aldehyde or keto end of the molecule projects to the right—are said to be of the D-configuration; those sugars derived from L-glyceraldehyde are said to be of the L-configuration.

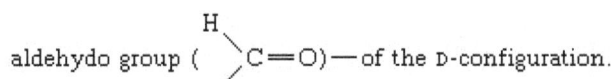

$$\text{aldehydo group } (\overset{H}{\diagdown}C=O) - \text{of the D-configuration.}$$

Representative Disaccharides and Oligosaccharides

common name	component sugars	linkages	sources
cellobiose	glucose, glucose	β1 → 4*	hydrolysis of cellulose
gentiobiose	glucose, glucose	β1 → 6	plant glycosides, amygdalin
isomaltose	glucose, glucose	α1 → 6	hydrolysis of glycogen, amylopectin
raffinose**	galactose, glucose, fructose	α1 → 6, α1 → 2	sugarcane, beets, seeds
stachyose**	galactose, galactose, glucose, fructose	α1 → 6, α1 → 6, α1 → 2	soybeans, jasmine, twigs, lentils

The linkage joins carbon atom 1 (in the β configuration) of one glucose molecule and carbon atom 4 of the second glucose molecule; the linkage may also be abbreviated β-1, 4.

The configurational notation D or L is independent of the sign of the optical rotation of a sugar in solution. It is common, therefore, to designate both, as, for example, D-(l)-fructose or D-(d)-glucose; i.e., both have a D-configuration at the centre of asymmetry most remote from the aldehyde end (in glucose) or keto end (in fructose) of the molecule, but fructose is levorotatory and glucose is dextrorotatory—hence the latter has been given the alternative name dextrose. Although the initial assignments of configuration for the glyceraldehydes were made on purely arbitrary grounds, studies that were carried out nearly half a century later established them as correct in an absolute spatial sense. In biological systems, only the D or L form may be utilized.

When more than one asymmetrical centre is present in a molecule, as is the case with sugars having four or more carbon atoms, a series of DL pairs exists, and they are functionally, physically, and chemically distinct. Thus, although D-xylose and D-lyxose both have five carbon atoms and are of the D-configuration, the spatial arrangement of the asymmetrical centres (at carbon atoms 2, 3, and 4) is such that they are not mirror images.

Hemiacetal and Hemiketal Forms

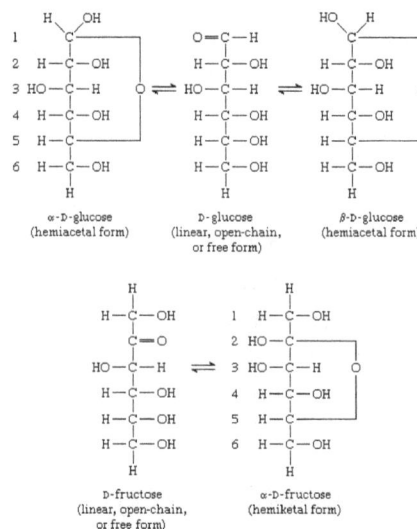

α-D-glucose (hemiacetal form)   D-glucose (linear, open-chain, or free form)   β-D-glucose (hemiacetal form)

D-fructose (linear, open-chain, or free form)   α-D-fructose (hemiketal form)

Although optical rotation has been one of the most frequently determined characteristics of carbohydrates since its recognition in the late 19th century, the rotational behaviour of freshly prepared solutions of many sugars differs from that of solutions that have been allowed to stand. This phenomenon, known as mutarotation, is demonstrable even with apparently identical sugars and is caused by a type of stereoisomerism involving formation of an asymmetrical centre at the first carbon atom (aldehyde carbon) in aldoses and the second one (keto carbon) in ketoses.

Most pentose and hexose sugars, therefore, do not exist as linear, or open-chain, structures in solution but form cyclic, or ring, structures in hemiacetal or hemiketal forms, respectively. As illustrated for glucose and fructose, the cyclic structures are formed by the addition of the hydroxyl group (−OH) from either the fourth, fifth, or sixth carbon atom to the carbonyl group C=O at position 1 in glucose or 2 in fructose. In the case of five-membered cyclic ketohexose or six-membered cyclic aldohexose, the cyclic forms are in equilibrium with (i.e., the rate of conversion from one form to another is stable) the open-chain structure—a free aldehyde if the solution contains glucose, a free ketone if it contains fructose; each form has a different optical rotation value. Since the forms are in equilibrium with each other, a constant value of optical rotation is measurable; the two cyclic forms represent more than 99.9 percent of the sugar in the case of a glucose solution.

The carbon atom containing the aldehyde or keto group is called the anomeric carbon atom; similarly, carbohydrate stereoisomers that differ in configuration only at this carbon atom are called anomers. When a cyclic hemiacetal or hemiketal structure forms, the structure with the new hydroxyl group projecting on the same side as that of the oxygeninvolved in forming the ring is called the alpha anomer; that with the hydroxyl group projecting on the opposite side from that of the oxygen ring is called the beta anomer.

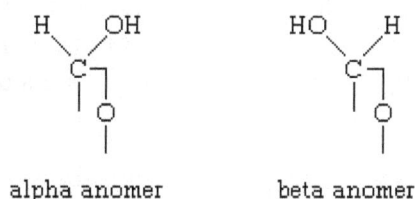

alpha anomer   beta anomer

The spatial arrangements of the atoms in these cyclic structures are better shown (glucose is used as an example) in the representation devised by British organic chemist Sir Norman Haworth about 1930; they are still in widespread use. In the formulation the asterisk indicates the position of the anomeric carbon atom; the carbon atoms, except at position 6, usually are not labelled.

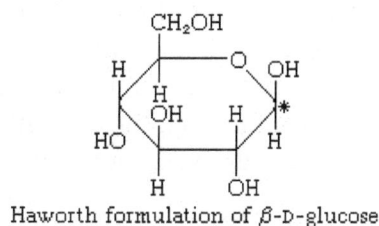

Haworth formulation of β-D-glucose

The large number of asymmetrical carbon atoms and the consequent number of possible isomers considerably complicates the structural chemistry of carbohydrates.

Monosaccharide

Monosaccharide, also called simple sugar is any of the basic compounds that serve as the building blocks of carbohydrates. Monosaccharides are polyhydroxy aldehydes or ketones; that is, they are molecules with more than one hydroxyl group ($-OH$), and a carbonyl group ($C=O$) either at the terminal carbon atom (aldose) or at the second carbon atom (ketose). The carbonyl group combines in aqueous solution with one hydroxyl group to form a cyclic compound (hemi-acetal or hemi-ketal). The resulting monosaccharide is a crystalline water-soluble solid.

Monosaccharides are classified by the number of carbon atoms in the molecule; dioses have two, trioses have three, tetroses four, pentoses five, hexoses six, and heptoses seven. Most contain five or six. The most-important pentoses include xylose, found combined as xylan in woody materials; arabinose from coniferous trees; ribose, a component of ribonucleic acids (RNA) and several vitamins; and deoxyribose, a component of deoxyribonucleic acid (DNA). Among the most-important aldohexoses are glucose, mannose, and galactose; fructose is a ketohexose.

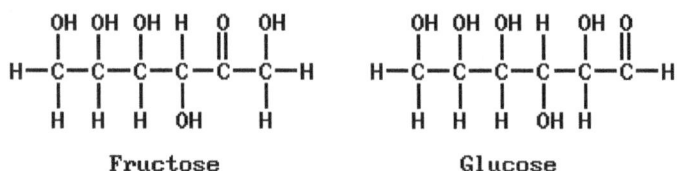

$$\begin{array}{ccccccc} OH & OH & OH & H & O & OH \\ | & | & | & | & || & | \\ H-C & -C & -C & -C & -C & -C-H \\ | & | & | & | & & | \\ H & H & H & OH & & H \end{array} \qquad \begin{array}{ccccccc} OH & OH & OH & H & OH & O \\ | & | & | & | & | & || \\ H-C & -C & -C & -C & -C & -C-H \\ | & | & | & | & | & \\ H & H & H & OH & H & \end{array}$$

Fructose   Glucose

Several derivatives of monosaccharides are important. Ascorbic acid (vitamin C) is derived from glucose. Important sugar alcohols (alditols), formed by the reduction of (i.e., addition of hydrogen to) a monosaccharide, include sorbitol (glucitol) from glucose and mannitol from mannose; both are used as sweetening agents. Glycosides derived from monosaccharides are widespread in nature, especially in plants. Amino sugars (i.e., sugars in which one or two hydroxyl groups are replaced with an amino group, $-NH_2$) occur as components of glycolipidsand in the chitin of arthropods.

Structure of Monosaccharide

Simple sugars can be defined as polyhydroxy-aldehydes or ketones. Hence the simplest sugars contain at least three carbons. The most common are the aldo- and keto-trioses, tetroses, pentoses, and hexoses. The simplest 3C sugars are glyceraldehye and dihydroxyacetone.

Glucose, an aldo-hexose, is a central sugar in metabolism. It and other 5 and 6C sugars can cyclize through intramolecular nucleophilic attack of one of the OH's on the carbonyl C of the aldehyde or ketone. Such intramolecular reactions occur if stable 5 or 6 member rings can form. The resulting rings are labeled furanose (5 member) or pyranose (6 member) based on their similarity to furan and pyran. On nucleophilic attack to form the ring, the carbonyl O becomes an OH which points either below the ring (a anomer) or above the ring (b anomer).

Monosaccharides in solution exist as equilibrium mixtures of the straight and cyclic forms. In solution, glucose is mostly in the pyranose form, fructose is 67% pyranose and 33% furanose, and ribose is 75% furanose and 25% pyranose. However, in polysaccharides, is exclusively pyranose and fructose and ribose are furanoses.

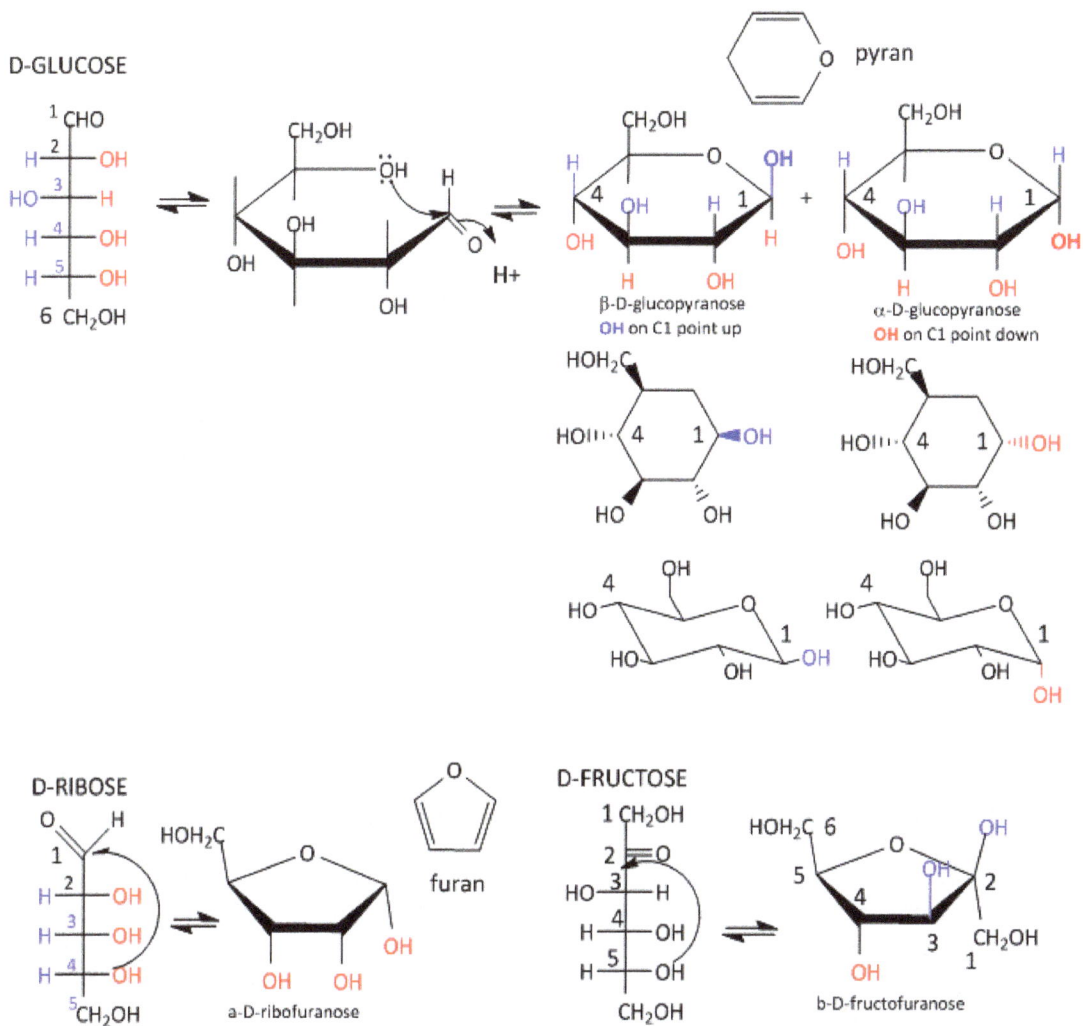

Figure: Sugar Ring Formation and Representations

Sugars can be drawn in the straight chain form as either Fisher projections or perspective structural formulas.

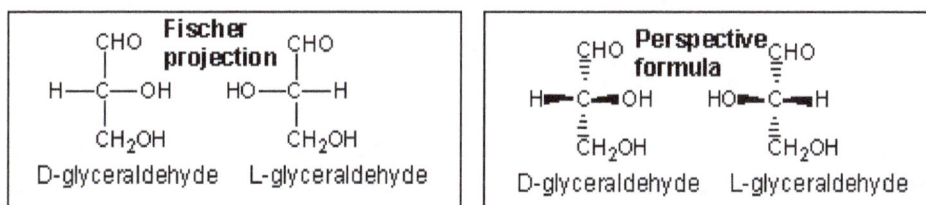

In the Fisher projection, the vertical bonds point down into the plane of the paper. That's easy to visualize for 3C molecules. but more complicated for bigger molecules. For those draw a wedge and dash line drawing of the molecule. When determining the orientation of the OHs on each C, orient the wedge and dash drawing in your mind so that the C atoms adjacent to the one of interest are pointing down. Sighting towards the carbonyl C, if the OH is pointing to the right in the Fisher project, it should be pointing to the right in the wedge and dash drawing, as shown below for D-erthyrose and D-glucose.

Fisher Projections

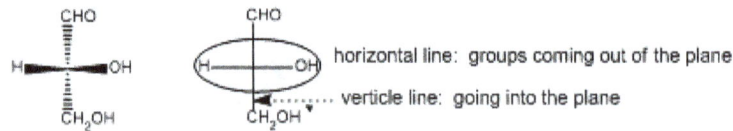

horizontal line: groups coming out of the plane

verticle line: going into the plane

D-glyceraldehyde

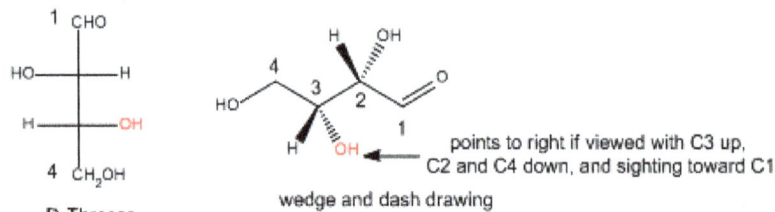

D-Threose

points to right if viewed with C3 up,
C2 and C4 down, and sighting toward C1

wedge and dash drawing

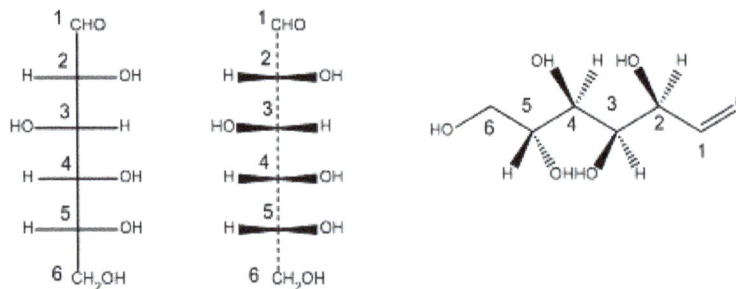

Figure: Orienting OH groups in wedge and dashing drawings of simple straight chain sugars

Cyclic forms can be drawn either as the Haworth projections, which shows the molecule as cyclic and planar with substituents above or below the ring) or the more plausible bent forms (showing Glc in the chair or boat conformations, for example). b-D-glucopyranose is the only aldohexose which can be drawn with all its bulky substituents (OH and CH2OH) in equatorial positions, which probably accounts for its widespread prevalence in nature.

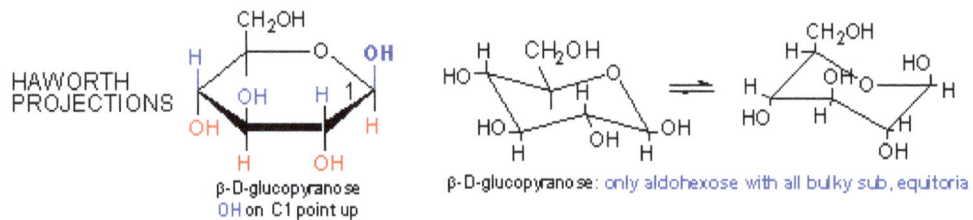

HAWORTH
PROJECTIONS

β-D-glucopyranose
OH on C1 point up

β-D-glucopyranose: only aldohexose with all bulky sub, equatoria

Haworth projections are more realistic than the Fisher projections, but you should be able to draw both structures. In general, if a substituent points to the right in the Fisher structure, it points down in the Haworth. if it points left, it points up. In general, the OH on the a-anomer points down (ants down) while on the b-anomer it points up (butterflies up).

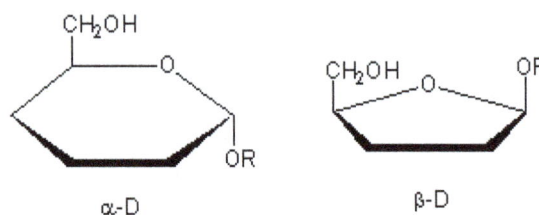

α-D   β-D

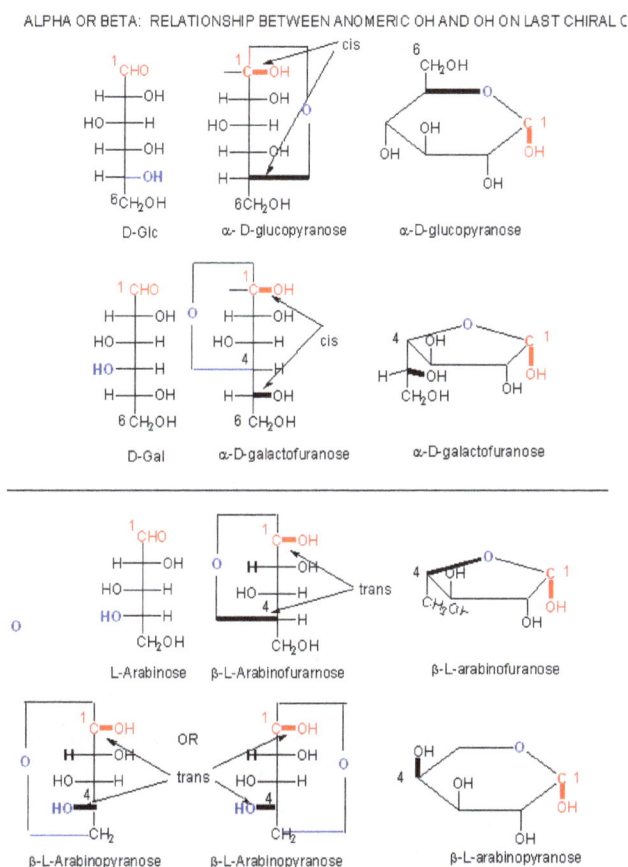

Figure: A more rigorous view of the relationship between the anomeric OH and the OH on the last chiral C of a sugar.

In the Haworth projections, the bulky R group of the next carbon after the carbon whose OH group engaged in a nucleop hilic attach on the carbonyl carbon to form the ring O is pointed up if the OH engaged in the attach was on the right hand side in the straight chain Fisher diagram (as in a-D-glucopyranose above when the CH_2OH group is up) but is pointed down if the OH engaged in the attach was on the left hand side in the straight chain Fisher diagram (as in a-D-galactofuranose above when the (CHOH)CH2OH group is down). The rest of the OH groups still follow the simple rule that if they are pointing to the right in the Fisher straight chain form, they point down in the Haworth form.

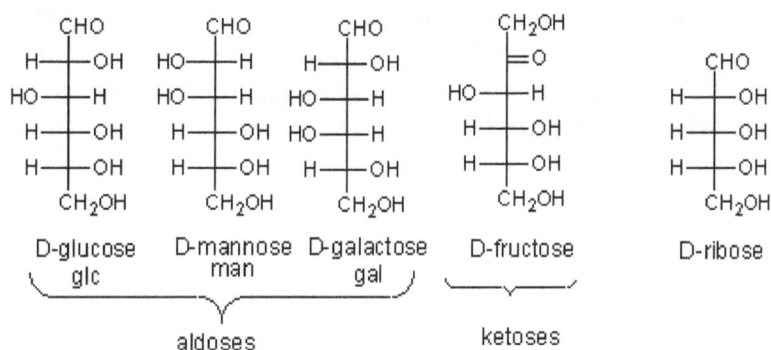

The most common monosaccharides (other than glyceraldehyde and dihydroxyacetone) which you need to know are shown below.

The mirror image of D-Glc is L-Glc. For common sugars, the prefix D and L refer to the center of asymmetry most remote from the aldehyde or ketone. By convention, all chiral centers are related to D- glyceraldehyde, so sugar isomers related to D-glyceraldehyde at their last asymmetric center are D sugars.

Function of Monosaccharide

Fuel for Metabolism

One major function of a monosaccharide is its use for energy within a living organism. Glucose is a commonly known carbohydrate that is metabolized within cells to create fuel. In the presence of oxygen, glucose breaks down into carbon dioxide and water, and energy is released as a by-product. Glucose is a product of photosynthesis, and plants obtain energy from glucose through respiration. Humans acquire glucose from food, and the body transforms this monosaccharide into energy.

Building Blocks

Monosaccharides are also the foundation for more complex carbohydrates, or they serve as components to amino acids. The ribose and deoxyribose monosaccharides are vital elements of RNA and DNA, which are the building blocks of life. While monosaccharides cannot be broken down into smaller sugars, disaccharides and polysaccharides are broken down into monosaccharides in processes like digestion. For example, the disaccharide lactose is degraded into monosaccharides, which can be absorbed into the human body.

Examples of Monosaccharides

Glucose

Glucose (Glc) is a monosaccharide (or simple sugar) with the chemical formula $C_6H_{12}O_6$. It is the major free sugar circulating in the blood of higher animals, and the preferred fuel of the brain and nervous system, as well as red blood cells (erythrocytes).

As a universal substrate (a molecule upon which an enzyme acts) for the production of cellular energy, glucose is of central importance in the metabolism of all life forms. It is one of the main products of photosynthesis, the process by which photoautotrophs such as plants and algae convert energy from sunlight into potential chemical energy to be used by the cell. Glucose is also a major starting point for cellular respiration, in which the chemical bonds of energy-rich molecules such as glucose are converted into energy usable for life processes.

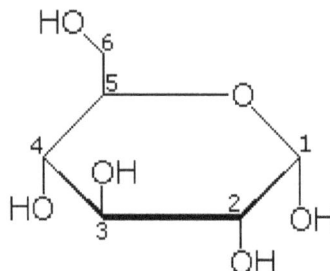

Chemical name	6-(hydroxymethyl)oxane-2,3,4,5-tetrol
Synonym for D-glucose	dextrose
Varieties of D-glucose	α-D-glucose; β-D-glucose
Abbreviations	**Glc**
Chemical formula	$C_6H_{12}O_6$
Molecular mass	180.16 g mol^{-1}
Melting point	α-D-glucose: 146 °C β-D-glucose: 150 °C
Density	1.54 g/cm^{-3}
CAS number	50-99-7 (D-glucose)
CAS number	921-60-8 (L-glucose)
SMILES	C(C1C(C(C(C(O1)O)O)O)O)O

Glucose stands out as a striking example of the complex interconnectedness of plants and animals: the plant captures solar energy into a glucose molecule, converts it to a more complex form(starch or cellulose) that is eaten by animals, which recover the original glucose units, deliver it to their cells, and eventually use that stored solar energy for their own metabolism. Milk cows, for example, graze on grass as a source of cellulose, which they break down to glucose using their four-chambered stomachs. Some of that glucose then goes into the milk we drink.

All major dietary carbohydrates contain glucose, either as their only building block, or in combination with another monosaccharide, as in sucrose ("table sugar") and lactose, the primary sugar found in milk.

The natural form of glucose (D-glucose) is also referred to as dextrose, especially in the food industry.

Structure of Glucose

Sugars are classified according to two properties: (1) number of carbon atoms and (2) type of functional group (either an aldehyde or a ketone group). Glucose, which has six carbon atoms (i.e., it is a hexose sugar) and contains an aldehyde group (-CHO), is thus referred to as an *aldohexose*.

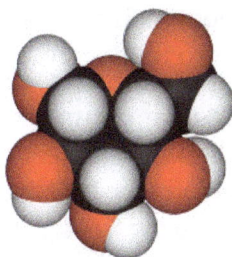

A space-filling model of glucose

```
       H         O
        \      //
         \    /
          \  /
           ||
    H ——————— OH

   HO ——————— H

    H ——————— OH

    H ——————— OH

         CHOH₂
```

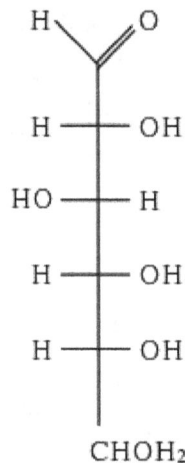

The open-chain form of D-glucose.

The glucose molecule can exist in an open-chain (acyclic) form and a ring (cyclic) form. In solution and at neutral pH, the cyclic form is predominant at equilibrium. When glucose exists in cyclic form, the functional group is not free, making the molecule less reactive. This preference for the less reactive ring form offers a possible explanation for the crucial and widespread use of glucose in metabolism, as opposed to another monosaccharide such as fructose (Fru). The low tendency of glucose, in comparison to other hexose sugars, to non-specifically react with the amino groups of proteins might explain its importance to advanced life.

Isomers

Glucose has four optic centers, which means that in theory glucose can have $(4^2-1) = 15$ optical stereoisomers. Only seven of these are found in living organisms, and of these galactose (Gal) and mannose (Man) are the most important. These eight isomers (including glucose itself) are all diastereoisomers in relation to each other (i.e., they are not mirror images), and all belong to the D-series.

Natural Sources of Glucose

- The conversion of light energy into chemical energy. Glucose is one of the products of photosynthesis in plants and algae, as well as some bacteria and protists.

- The breakdown of storage forms of glucose. Glucose can be obtained through the breakdown of glycogen, the storage form of glucose in animals and fungi, through a process known as glycogenolysis. Glycogen is an auxiliary energy source, tapped and converted back into glucose when there is need for energy. In plants, glucose is stored as starch.

- The synthesis of glucose from non-carbohydrates. When glucose is not supplied in the diet and glycogen stores have been depleted, animals may also synthesize glucose in the liver and (to a lesser extent) in the kidneys from non-carbohydrate intermediates. Lactate from active skeletal muscle, amino acids from protein in the diet or protein in muscle, and glycerol, derived from the breakdown of fats, may contribute to the synthesis of glucose (gluconeogenesis).

Functions of Glucose

Because the cell membrane is permeable to glucose, the cell cannot accumulate pure glucose to any higher concentration than is present in the bloodstream. Cells, nonetheless, do accumulate glucose as an enzyme chemically modifies the glucose molecule by the addition of a phosphate group (phosphorylation). Since the cell membrane is impermeable to this modified form, called glucose-6-phosphate, the process effectively "traps" glucose inside the cell, allowing the recovery of more glucose from the bloodstream. Glucose-6-phosphate, in turn, can be used for three major functions, depending on the specific conditions within the cell and the overall needs of the organism:

Glucose is a Major Energy Source

Glucose is a ubiquitous fuel in biology. When chemical energy is needed, glucose is oxidized to pyruvate through a process known as glycolysis, which is the energy source for certain organisms called obligate anaerobes that cannot utilize oxygen for metabolism. In aerobic organisms, however, the pyruvate typically continues onward to the reactions of the Citric acid cycle (TCAC) and the electron transport chain, forming CO_2 and water. These later reactions generate about 18 times more energy than glycolysis, mostly in the form of ATP.

Glucose Plays a Role in the Synthesis of Non-carbohydrates

Glucose and its metabolites can also be mobilized when carbon skeletons are needed. That is, glucose also participates in the synthesis of complex molecules (anabolism) in addition to its role in the catabolic pathways that break down molecules into smaller components. For example, glucose-6-phosphate can enter the pentose phosphate pathway, which generates the five-carbon (pentose) sugar ribose for the synthesis of nucleotides, the building blocks of the nucleic acids DNA and RNA.

In plants and most animals (excepting guinea pigs and primates, such as humans), glucose is a precursor for the production of vitamin C (ascorbic acid). Polymers of glucose may also be bound to proteins (to form glycoproteins) or lipids (to form glycolipids). The addition of sugar chains may function to assist proteins in folding into their characteristic three-dimensional structure, to enhance the stability of proteins and membrane lipids, or to act as recognition sites for specific chemicals.

Glucose is a Component of other Carbohydrates

When the organism has an abundant supply of ATP and glucose, then it can synthesize one or more of the common glucose polymers (polysaccharides): glycogen for animals and starch and cellulose for plants. While glycogen and starch serve as energy-storage molecules, cellulose plays a primarily structural role in green plants.

Regulation of Blood Glucose

Given the importance of glucose as the preferred fuel of the brain, a constant blood glucose level (which typically falls between 4.4 mM and 6.7 mM in an adult male human) must be maintained for health and survival.

The concentration of blood glucose is mainly regulated through the action of hormones. Specifically, the hormone insulin directs the flow of glucose from the blood into liver, muscle, and adipose (fat storage) cells. It also promotes the increased synthesis of glycogen when energy needs have been met (for example, after a meal). When blood glucose levels fall (e.g., several hours after a meal), glucagon and epinephrine (also known as adrenaline) retrieve glucose from its storage form as glycogen in liver and muscle tissue. Low levels of insulin in this state also mean that the entry of glucose into muscle and adipose cells decreases, so that these cells switch to the use of non-carbohydrate fuels.

The liver is a major control site of blood glucose levels, with the ability to respond to hormonal signals that indicate either reduced or elevated blood glucose levels. One of the most important functions of the liver is to produce glucose for circulation.

Low blood glucose levels (hypoglycemia) can result in impaired functioning of the central nervous system, which may manifest itself in dizziness, speech problems, or even loss of consciousness.

Hyperglycemia (elevated blood sugar), which is characteristic of diabetes mellitus, indicates an overproduction of glucose by the liver cells accompanied by an inability of other cells to utilize glucose. Patients with type 1 diabetes mellitus depend on external sources of insulin for their survival because (in most cases) their autoimmune system destroys the cells in the pancreas that secrete insulin. Patients with the more common type 2 diabetes mellitus may have relatively low insulin production or resistance to its effects.

Commercial Production

Glucose is produced commercially through the breakdown of starch in an enzyme-catalyzed process called hydrolysis (a chemical reaction in which a molecule is split into two parts through the addition of water). The enzymatic process has two stages:

- Over the course of one to two hours near 100 °C, enzymes break the starch into smaller carbohydrates containing on average 5-10 glucose units each.

- In the second step, known as saccharification, the partially hydrolyzed starch is completely hydrolyzed to glucose using the glucoamylase enzyme from the fungus Aspergillus niger. Typical reaction conditions are pH 4.0–4.5, 60 °C, and a carbohydrate concentration of 30–35 percent by weight. Under these conditions, starch can be converted to glucose at 96 percent yield after one to four days.

The resulting glucose solution is then purified by filtration and concentrated in a multiple-effect evaporator. Solid D-glucose is finally produced by repeated crystallizations.

Fructose

Fructose is is a simple sugar (monosaccharide) with the same chemical formula as glucose ($C_6H_{12}O_6$) but a different atomic arrangement. Along with glucose and galactose, fructose is one of the three most important blood sugars in animals.

Sources of fructose include honey, fruits, and some root vegetables. Fructose is often found in combination with glucose as the disaccharide sucrose (table sugar), a readily transportable and mobilizable sugar that is stored in the cells of many plants, such as sugar beets and sugarcane. In animals, fructose may also be utilized as an energy source, and phosphate derivatives of fructose participate in carbohydrate metabolism.

In addition to natural sources, fructose may be found in commercially produced high fructose corn syrup (HFCS). Like regular corn syrup, HFCS is derived from the hydrolysis of corn starch to yield glucose; however, further enzymatic processing occurs to increase the fructose content. Until recently, fructose has not been present in large amounts in the human diet; thus, the increasing consumption of HFCS as a sweetener in soft drinks and processed foods has been linked to concerns over the rise in obesity and type II diabetes in the United States.

Fructose's Glycemic Index (an expression of the relative ability of various carbohydrates to raise blood glucose level) is relatively low compared to other simple sugars. Thus, fructose may be recommended for persons with diabetes mellitus or hypoglycemia (low blood sugar), because intake does not trigger high levels of insulin secretion. This benefit is tempered by a concern that fructose may have an adverse effect on plasma lipid and uric acid levels, and that higher blood levels of fructose can be damaging to proteins.

Chemical Structure of Fructose

Fructose is a levorotatory monosaccharide (counterclockwise rotation of plane polarized light) with the same empirical formula as glucose but with a different structural arrangement of atoms (i.e., it is an isomer of glucose). Like glucose, fructose is a hexose (six-carbon) sugar, but it contains a keto group instead of an aldehyde group, making it a ketohexose.

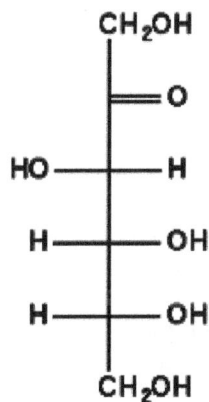

$$
\begin{array}{c}
CH_2OH \\
| \\
{=}O \\
| \\
HO{-}\!\!\!{-}H \\
| \\
H{-}\!\!\!{-}OH \\
| \\
H{-}\!\!\!{-}OH \\
| \\
CH_2OH
\end{array}
$$

The open-chain structure of fructose

Like glucose, fructose can also exist in ring form. Its open-chain structure is able to cyclize (form a ring structure) because a ketone can react with an alcohol to form a hemiketal. Specifically, the C-2 keto group of a fructose molecule can react with its C-5 hydroxyl group to form an intramolecular hemiketal. Thus, although fructose is a hexose, it may form a five-membered ring called a furanose, which is the structure that predominates in solution.

Fructose's specific conformation (or structure) is responsible for its unique physical and chemical properties relative to glucose. For example, although the perception of sweetness depends on a variety of factors, such as concentration, pH, temperature, and individual taste buds, fructose is estimated to be approximately 1.2-1.8 times sweeter than glucose.

Fructose as an Energy Source

Fructose Absorption

Fructose is absorbed more slowly than glucose and galactose, through a process of facilitated diffusion (in which transport across biological membranes is assisted by transport proteins). Large amounts of fructose may overload the absorption capacity of the small intestine, resulting in diarrhea. For example, young children who drink a lot of fruit juice that is composed mainly of fructose may suffer from "toddlers' diarrhea." Fructose is absorbed more successfully when ingested with glucose, either separately or as sucrose.

Most dietary fructose is then metabolized by the liver, a control point for the circulation of blood sugar.

Breakdown of Fructose

Energy from carbohydrates is obtained by nearly all organisms via glycolysis. It is only the initial stage of carbohydrate catabolism for aerobic organisms such as humans. The end-products of glycolysis typically enter into the citric acid cycle and the electron transport chain for further oxidation, producing considerably more energy per glucose molecule.

Fructose may enter the glycolytic pathway by two major routes: one predominant in liver, the other in adipose tissue (a specialized fat-storage tissue) and skeletal muscle. In the latter, the degradation of fructose closely resembles the catabolism of glucose: the enzyme hexokinase phosphorylates (adds a phosphate) to form fructose-6-phosphate, an intermediate of glycolysis.

The liver, in contrast, handles glucose and fructose differently. There are three steps involved in the fructose-1-phosphate pathway, which is preferred by liver due to its high concentration of fructokinase relative to hexokinase:

1. Fructose is phosphorylated by the enzyme fructokinase to fructose-1-phosphate.

2. The six-carbon fructose is split into two three-carbon molecules, glyceraldehyde and dihydroxyacetone phosphate.

3. Glyceraldehyde is then phosphorylated by another enzyme so that it too can enter the glycolytic pathway.

Galactose

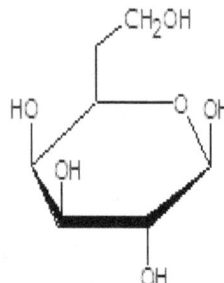

Figure: beta-D-Galactopyranose

Galactose is a monosaccharide and the C4 epimer of glucose, that is, they differ only for the position of the -OH group on C4 (axial in Gal, equatorial in glucose).

It has a sweetness equal to 33% of sucrose.

Food Sources of Galactose

In human nutrition the most part comes from the hydrolysis of the disaccharide lactose, the milk sugar, including that of the human milk.

As mother's milk is the only source of energy and carbohydrates for newborn, galactose has a crucial role in human nutrition.

The monosaccharide is also bound to caseins, and therefore it is found in all dairy products.

Findings from studies conducted since the 50s of last century have shown that galactose is present not only in milk and dairy products, but also in plant products such as legumes, grains, nuts, tubers and vegetables. It seems that in these foods it is often engaged in bonds resistant to the attack of human digestive enzymes, and therefore not metabolizable. However, different fruits and vegetables contain it also in free form, in variable amounts:

- less than 0.1 mg/100 g of edible portion: artichokes, mushrooms, olives, and peanuts;

- more than 10 mg/100 g of edible portion: bell peppers, date, papaya, watermelon, tomato;

- up to 35.4 mg/100 g of edible portion in persimmon.

These values are very low, but to take into account in case of galactosemia.

Metabolism

Lactose hydrolysis by intestinal lactase (EC 3.2.1.108) leads to galactose release, together a glucose molecule. It is a β-(1→4)-glycosidic reaction. Lactase activity is present in a multifunctional enzyme containing also an active site capable of hydrolyzing milk glycolipids, namely ceramides to yield fatty acids and sphingosine (EC 3.2.1.62).

Also bacterial β-galactosidase (EC 3.2.1.23) in yogurt is able to convert milk sugar into its constituent monosaccharides.

Free galactose is then absorbed through the mucosa of the small intestine, passes into the portal circulation and is transported to the liver, where it is almost completely absorbed, so that its blood concentration does not exceed 1 mmol/L. It should be noted that glucose represents over 95% of hexoses found in the blood stream.

Under normal physiological conditions one can observe an increase in its blood concentration as a result of alcohol consumption, which reduces its intestinal absorptionbut also the subsequent hepatic metabolism.

Leloir Pathway

In hepatocytes, galactose enters the Leloir pathway.

Freed from lactose, the monosaccharide is mostly present as beta-isomer, and the first step of its hepatic metabolism is the conversion to the alpha-isomer, in the reaction catalyzed by galactose mutarotase, also known as aldose 1-epimerase.

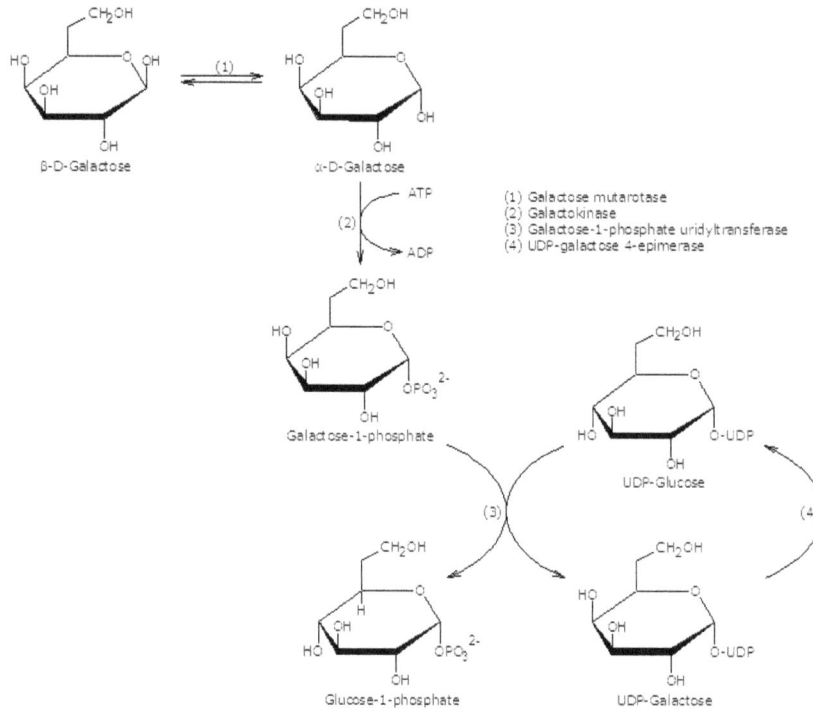

Figure: Leloir Pathway

In the second step, phosphorylation occurs of alpha-D-Gal to Gal-1-phosphate, in the reaction catalyzed by galactokinase (EC 2.7.1.6), a phosphorylation at C-1.

In the next step, galactose-1-phosphate uridyltransferase or GALT (EC 2.7.7.12) catalyzes the transfer of a UMD group from UDP-Gal to glucose-1-phosphate, with formation of glucose-1-phosphate and UDP-Gal.

The cycle ends when the UDP-Gal is converted to UDP-glucose in the reaction catalyzed by UDP-galactose 4-epimerase or GALE (EC 5.1.3.2).

Although in theory the glucose-1-phosphate product may be converted into glucose-6-phosphate in the reaction catalyzed by phosphoglucomutase (EC 5.4.2.2), and then enter the glycolytic pathway, it seems that only a small part of the ingested galactose follows this pathway. Conversely, glucose-1-phosphate, activated to UDP-glucose, is used for glycogen synthesis.

Metabolic Fate of UDP-Gal

- UDP-Gal is an important precursor in the synthesis of glycolipids, such as gangliosides and galactocerebrosides, sphingolipids, mucopolysaccharides, and membrane glycoproteins.

- In the adult mammary gland, under the influence of prolactin, UDP-Gal can be joined to glucose to give milk sugar.

Galactosemia

Mutations in three of the four enzymes of the Leloir pathway, i.e. galactokinase, GALT or GALE, that cause their malfunction, lead to galactosemia, a pathological condition less frequent but more severe than lactose intolerance.

In the disease, there is an increase in the blood concentration of galactose to values higher than 1 mmol/L; different tissues remove it from the blood stream and reduce it to galactitol (dulcitol) in the reaction catalyzed by aldehyde reductase. Galactitol is not further metabolized, accumulates in tissues and causes pathological changes resulting from the increase in osmotic pressure caused by it.

Galactose and Myelin

Myelin is the covering sheath of axons of neurons, where it plays an insulating and protective role, crucial for the conduction of nerve impulses. Lipids account for about 70-80% of the dry weight of myelin, proteins 20-30%. In lipid fraction, in addition to cholesterol and phosphoglycerides, galactocerebrosides are also found.

The participation in the formation of the myelin sheath of nerve fibers, that begins during fetal life and is completed at second childhood, is by far the most important function of galactose.

Disaccharide

A disaccharide, also called a double sugar, is a molecule formed by two monosaccharides, or simple sugars. Three common disaccharides are sucrose, maltose, and lactose. They have 12 carbon atoms, and their chemical formula is $C_{12}H_{22}O_{11}$. Other, less common disaccharides include lactulose, trehalose, and cellobiose. Disaccharides are formed through dehydration reactions in which a total of one water molecule is removed from the two monosaccharides.

Functions of Disaccharides

Disaccharides are carbohydrates found in many foods and are often added as sweeteners. Sucrose, for example, is table sugar, and it is the most common disaccharide that humans eat. It is also found in other foods like beetroot. When disaccharides like sucrose are digested, they are broken down into their simple sugars and used for energy. Lactose is found in breast milk and provides nutrition for infants. Maltose is a sweetener that is often found in chocolates and other candies.

Plants store energy in the form of disaccharides like sucrose and it is also used for transporting nutrients in the phloem. Since it is an energy storage source, many plants such as sugar cane are high in sucrose. Trehalose is used for transport in some algae and fungi. Plants also store energy in polysaccharides, which are many monosaccharides put together. Starch is the most common polysaccharide used for storage in plants, and it is broken down into maltose. Plants also use disaccharides to transport monosaccharides like glucose, fructose, and galactose between cells.

Packaging monosaccharides into disaccharides makes the molecules less likely to break down during transport.

Formation and Breakdown of Disaccharides

When disaccharides are formed from monosaccharides, an -OH (hydroxyl) group is removed from one molecule and an H (hydrogen) is removed from the other. Glycosidic bonds are formed to join the molecules; these are covalent bonds between a carbohydrate molecule and another group (which does not necessarily need to be another carbohydrate). The H and -OH that were removed from the two monosaccharides join together to form a water molecule, H_2O. For this reason, the process of forming a disaccharide from two monosaccharides is called a dehydration reaction or condensation reaction.

When disaccharides are broken down into their monosaccharide components via enzymes, a water molecule is added. This process is called hydrolysis. It should not be confused with the process of dissolution, which happens when sugar is dissolved in water, for example. The sugar molecules themselves do not change structure when they are dissolved. The solid sugar simply turns into liquid and becomes a solute, or a dissolved component of a solution.

Uses of Disaccharides

Disaccharides are used as energy carriers and to efficiently transport monosaccharides. Specific examples of uses include:

- In the human body and in other animals, sucrose is digested and broken into its component simple sugars for quick energy. Excess sucrose can be converted from a carbohydrate into a lipid for storage as fat. Sucrose has a sweet flavor.

- Lactose (milk sugar) is found in human breast milk, where it serves as a chemical energy source for infants. Lactose, like sucrose, has a sweet flavor. As humans age, lactose becomes less-tolerated. This is because lactose digestion requires the enyzme lactase. People who are lactose intolerant can take a lactase supplement to reduce bloating, cramping, nausea, and diarrhea.

- Plants use disaccharides to transport fructose, glucose, and galactose from one cell to another.

- Maltose, unlike some other disaccharides, does not serve a specific purpose in the human body. The sugar alcohol form of maltose is maltitol, which is used in sugar-free foods. Of course, maltose is a sugar, but it is incompletely digested and absorbed by the body (50 to 60 percent).

Structure of Disaccharide

A disaccharide molecule is formed by 2 monosaccharides, joined by a glycosidic bond. The type of a glycosidic bond can determine the properties of certain disaccharides. For example, sucrose, isomaltulose and trehalulose are all composed of glucose and fructose, which are linked by different types of glycosidic bonds.

A Disaccharide Example

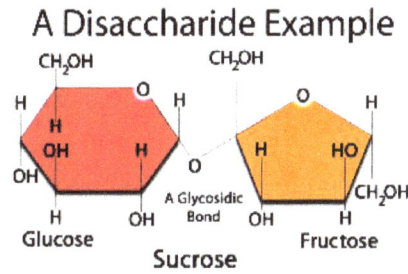

A disaccharide sucrose composed of monosaccharides glucose and fructose

Examples of Disaccharides

Sucrose

Sucrose, commonly known as "table sugar" or "cane sugar", is a carbohydrate formed from the combination of glucose and fructose. Glucose is the simple carbohydrate formed as a result of photosynthesis. Fructose is nearly identical, except for the location of a double-bonded oxygen. They are both six-carbon molecules, but fructose has a slightly different configuration. When the two combine, they become sucrose.

Plants use sucrose as a storage molecule. For quick energy, cells may store the sugar for later use. If far too much is accumulated, plants may begin to combine the complex sugars like sucrose into even large and denser molecules, like starches. These molecules, and oily lipids, are the main storage chemicals used by plants. In turn, animals eat these sugars and starches, break them back down into glucose, and use the energy within the bonds of glucose to power our cells.

Sucrose has been an important sugar for humans because it is easy extracted from plants such as sugar cane and sugar beets. These plants tend to store an excess of sugar, and from this we produce the majority of the sugar that we use. Even most "natural" sweeteners, which claim to be healthier than sucrose, are simply a different version of glucose combined in a different manner by plants.

Sucrose Structure

As mentioned above, sucrose is disaccharide, or a molecule made of two monosaccharides. Glucose and fructose are both monosaccharides, but together they make the disaccharide sucrose. This is an important process for the storage and compression of energy. Plants do this to make it easier to transport large amounts of energy, via sucrose. This process can be seen in the following image.

Glucose is seen on the left. Glucose is known as an aldose, meaning the carbonyl group (carbon double bonded to an oxygen) is found at the end of the chain of carbons. When the molecule creates a ring back on itself, it forms a 6-sided ring. Fructose, on the other hand, is a ketose. This means that the carbonyl group is found in the middle of the middle of the molecule. In this case, it forces fructose into a five-sided ring.

In a plant creating sucrose, an enzyme comes along to smash these two rings together, and extract a molecule of water. This process is called a condensation reaction, and forms a glycosidic bond between the two molecules. As you can see in the image, the reaction can also go the other way. To dissolve sucrose into fructose and glucose, a molecule of water can be added back in. This is what happens to sucrose as you digest it.

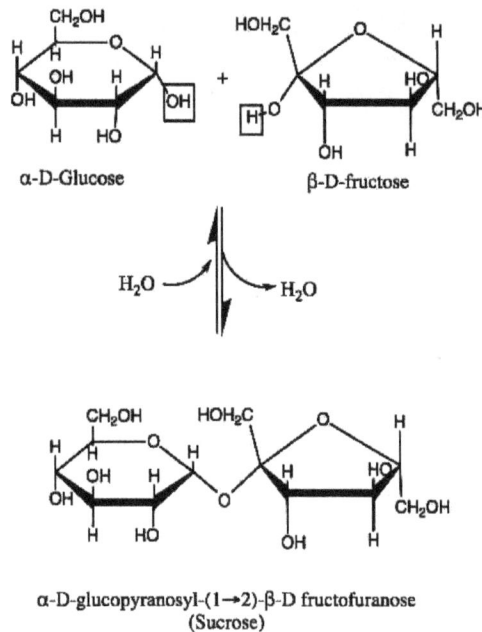

α-D-glucopyranosyl-(1→2)-β-D fructofuranose
(Sucrose)

Sucrose Uses

Sucrose is the most common form of carbohydrate used to transport carbon within a plant. Sucrose is able to be dissolved into water, while maintaining a stable structure. Sucrose can then be exported by plant cells into the phloem, the special vascular tissue designed to transport sugars. From the cells in which it was produces, the sucrose travels through the intercellular spaces within the leaf. It arrives at the vascular bundle, where specialized cells pump it into the phloem. The xylem, or vascular tube which carries water, adds small amounts of water to the phloem to keep the sugar mixture from solidifying. The sucrose mixture then makes its way down the phloem, arriving at cells in the stem and roots which have no chloroplasts and rely on the leaves for energy.

The sucrose is absorbed into these cells, and enzymes begin breaking the sucrose back into its constituent parts. The six-carbon glucose and fructose can be broken down into 3-carbon molecules, which are imported into the mitochondria, where they go through the citric acid cycle (AKA the Krebs Cycle). This process reduces coenzymes, which are then used in oxidative phosphorylation to create ATP. The energy within the bonds of ATP can power many of the reactions these cells need to complete in order to maintain the stem and roots.

Likewise, all other life on Earth is dependent upon sucrose and other carbs produced by plants. Sucrose was one of the first substances to be extracted from plants on a mass-scale, creating the white table sugar we know today. These sugars are extracted and purified from large crops, including sugar cane and sugar beets. To extract the sugar, the plants are usually boiled or heated, releasing the sugar. "Sugar in the Raw" is sugar which has not been treated further, while white table sugar undergoes more purification.

Lactose

Lactose is a disaccharide that consists of ß-D-galactose and ß-D-glucose molecules bonded through a ß1-4 glycosidic linkage. Lactose makes up around 2-8% of the solids in milk. The name comes from the Latin word for milk, plus the -ose ending used to name sugars. Its empirical formula is and its $C_{12}H_{22}O_{11}$ molecular weight is 342.3 g/mol.

Lactose

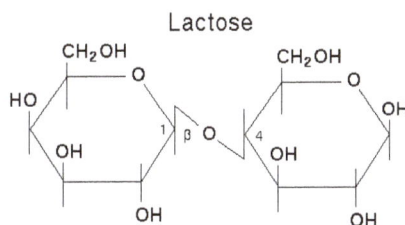

Digestion of Lactose

Infant mammals are fed on milk by their mothers. To digest it an enzyme called lactase (ß1-4 disaccharidase) is secreted by the intestinal villi, and this enzyme cleaves the molecule into its two subunits for absorption. Since lactose occurs mostly in milk, in most species the production of lactase gradually ceases with maturity, and they are then unable to metabolise lactose. This loss of lactase on maturation is also the default pattern in most adult humans. However, many people with ancestry in Europe, the Middle East, India, and the Maasai of East Africa, have a version of the gene for lactase that is not disabled after infancy, and in many of these cultures other mammals such as cattle, goats, and sheep are milked for food. This fact may cast doubt on some arguments by proponents of the Paleolithic diet, who argue that human metabolic needs have not changed since the last ice age. The process of retaining infant characteristics into adulthood is one of the simplest routes of adaptation, and is known as neoteny.

Structure and Reactions

Lactose is a disaccharide derived from the condensation of galactose and glucose, which form a β-1→4 glycosidic linkage. Its systematic name is β-D-galactopyranosyl-(1→4)-D-glucose. The glucose can be in either the α-pyranose form or the β-pyranose form, whereas the galactose can only have the β-pyranose form: hence α-lactose and β-lactose refer to the anomeric form of the glucopyranose ring alone.

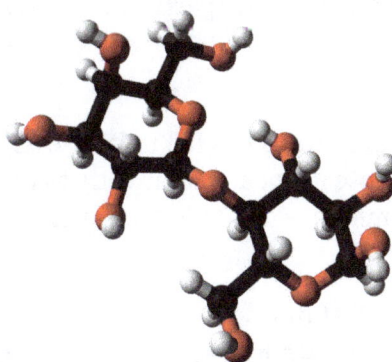

The molecular structure of α-lactose, as determined by X-ray crystallography.

Lactose is hydrolysed to glucose and galactose, isomerised in alkaline solution to lactulose, and catalytically hydrogenated to the corresponding polyhydric alcohol, lactitol. Lactulose is a commercial product, used for treatment of constipation.

Occurrence and Isolation

Lactose comprises about 2–8% of milk by weight. Several million tons are produced annually as a by-product of the dairy industry.

Whey or milk plasma is the liquid remaining after milk is curdled and strained, for example in the production of cheese. Whey is made up of 6.5% solids, of which 4.8% is lactose, which is purified by crystallisation. Industrially, lactose is produced from whey permeate – that is whey filtrated for all major proteins. The protein fraction is used in infant nutrition and sports nutrition while the permeate can be evaporated to 60–65% solids and crystallized while cooling. Lactose can also be isolated by dilution of whey with ethanol.

Dairy products such as yogurt, cream and fresh cheeses have lactose contents similar to that of milk. Ripened cheeses contain little to no lactose, as bacteria convert most of it into lactic acid during the ripening process.

Metabolism

Infant mammals nurse on their mothers to drink milk, which is rich in lactose. The intestinal villi secrete the enzyme lactase (β-D-galactosidase) to digest it. This enzyme cleaves the lactose molecule into its two subunits, the simple sugars glucose and galactose, which can be absorbed. Since lactose occurs mostly in milk, in most mammals, the production of lactase gradually decreases with maturity due to a lack of continuing consumption.

Many people with ancestry in Europe, West Asia, South Asia, the Sahel belt in West Africa, East Africa and a few other parts of Central Africa maintain lactase production into adulthood. In many of these areas, milk from mammals such as cattle, goats, and sheep is used as a large source of food. Hence, it was in these regions that genes for lifelong lactase production first evolved. The genes of adult lactose tolerance have evolved independently in various ethnic groups. By descent, more than 70% of western Europeans can drink milk as adults, compared with less than 30% of people from areas of Africa, eastern and south-eastern Asia and Oceania. In people who are lactose intolerant, lactose is not broken down and provides food for gas-producing gut flora, which can lead to diarrhea, bloating, flatulence, and other gastrointestinal symptoms.

Applications

Its mild flavor and easy handling properties has led to its use as a carrier and stabiliser of aromas and pharmaceutical products. Lactose is not added directly to many foods, because its solubility is less than that of other sugars commonly used in food. Infant formula is a notable exception, where the addition of lactose is necessary to match the composition of human milk.

Lactose is not fermented by most yeast during brewing, which may be used to advantage. For example, lactose may be used to sweeten stout beer; the resulting beer is usually called a milk stout or a cream stout.

Yeast belonging to the genus *Kluyveromyces* have a unique industrial application as they are capable of fermenting lactose for ethanol production. Surplus lactose from the whey by-product of dairy operations is a potential source of alternative energy.

Another significant lactose use is in the pharmaceutical industry. Lactose is added to tablet and capsule drug products as an ingredient because of its physical and functional properties, *i.e.*, compressibility and cost effective use. For similar reasons it can be used to cut (dilute) illicit drugs.

Maltose

Maltose, or malt sugar, is a disaccharide formed from two units of glucosejoined with an $\alpha(1{\to}4)$ linkage. Maltose is not common in food, but can be formed from the digestion of starch, and is heavy in the sugar in malt, the juice of barley and other grains. Maltose is a member of an important biochemical series of glucose chains. The disaccharides maltose, sucrose, and lactose have the same chemical formula, $C_{12}H_{22}O_{11}$, however, they differ in structure.

Maltose can be produced from starch by hydrolysis in the presence of the enzyme diastase. It can be broken down into two glucose molecules by hydrolysis. In living organisms, the enzyme maltase can achieve this very rapidly. In the laboratory, heating with a strong acid for several minutes will produce the same result.

There is another disaccharide that can be made from two glucoses, cellobiose, which differs only in the type of linkage used in the bond. While the difference between these two structures, maltose and cellobiose, is subtle, cellobiose has very different properties and cannot be hydrolized to glucose in the human body. The particular pathways used for these two molecules (some organisms can digest cellobiose) reflects the complex coordination in nature.

Maltose is important in the fermentation of alcohol, as starch is converted to carbohydrates and is readily broken down into glucose molecules with the maltase enzyme present in yeast. When cereals such as barley is malted, it is brought into a condition in which the concentration of maltose has been maximized. Metabolism of maltose by yeast during fermentation then leads to the production of ethanol and carbon dioxide.

Structure

Maltose is a carbohydrate (sugar). Carbohydrates are a class of biological molecules that contain primarily carbon (C) atoms flanked by hydrogen (H) atoms and hydroxyl (OH) groups (H-C-OH). They are named according to the number of carbon atoms they contain, with most sugars having between three and seven carbon atoms termed *triose* (three carbons), *tetrose* (four carbons), *pentose* (five carbons), *hexose* (six carbons), or *heptose* (seven carbons).

The single most common monosaccharide is the hexose D-glucose, represented by the formula $C_6H_{12}O_6$. In addition to occurring as a free monosaccharide, glucose also occurs in disaccharides, which consist of two monosaccharide units linked covalently. Each disaccharide is formed by a condensation reaction in which there is a loss of hydrogen (H) from one molecule and a hydroxyl group (OH) from the other. The resulting glycosidic bond—those that join a carbohydrate molecule to an alcohol, which may be another carbohydrate—is the characteristic linkage between sugars, whether between two glucose molecules, or between glucose and fructose, and so forth. When two glucose molecules are linked together, such as in maltose, glycosidic bonds form between carbon 1 of the first glucose molecule and carbon 4 of the second glucose molecule. (The carbons of glucose are numbered beginning with the more oxidized end of the molecule, the carbonyl group).

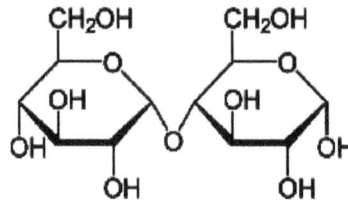

Maltose, pictured here, has an α-linkage, the OH group of carbon 1 on the first glucose points downwards. Cellobiose has a β-linkage, the OH group of carbon 1 on the first glucose points upwards.

Three common disaccharides are maltose, sucrose, and lactose. They share the same chemical formula, $C_{12}H_{22}O_{11}$, but involve different structures. Whereas *maltose* links two glucose units by an α(1→4) glycosidic linkage, *lactose* (milk sugar) involves glucose and galactose bonded through a β1-4 glycosidic linkage, and *sucrose* (common table sugar) consists of a glucose and a fructose joined by a glycosidic bond between carbon atom 1 of the glucose unit and carbon atom 2 of the fructose unit.

Although the disaccharide maltose contains two glucose molecules, it is not the only disaccharide that can be made from two glucoses. When glucose molecules form a glycosidic bond, the linkage will be one of two types, α or β, depending on whether the molecule that bonds its carbon 1 is an α-glucose or β-glucose. An α-linkage with carbon 4 of a second glucose molecule results in maltose, whereas a β-linkage results in *cellobiose*. As disaccharides, maltose and cellobiose also share the same formula $C_{12}H_{22}O_{11}$, but they are different compounds with different properties. For example, maltose can be hydrolyzed to its monosaccharides in the human body where as cellobiose cannot. Some organisms have the capacity to break down cellobiose.

The addition of another glucose unit yields maltotriose. Further additions will produce dextrins, also called maltodextrins, and eventually starch.

Function

Maltose is an important intermediate in the digestion of starch. Starch is used by plants as a way to store glucose. After cellulose, starch is the most abundant polysaccharide in plant cells. Animals (and plants) digest starch, converting it to glucose to serve as a source of energy. Maltose can form from this starch when it is broken down, and it in turn can be readily digested into the glucose molecules, the major free sugar circulating in the blood of higher animals, and the preferred fuel of the brain and nervous system.

Maltose is an interesting compound because of its use in alcohol production. Through a process called fermentation, glucose, maltose, and other sugars are converted to ethanol by yeast cells in the absence of oxygen. Through an analogous process, muscle cells convert glucose into lactic acid to obtain energy while the body operates under anaerobic conditions. Although maltose is uncommon in nature, it can be formed through the breakdown of starch by the enzymes of the mouth.

Polysaccharide

A polysaccharide is a large molecule made of many smaller monosaccharides. Monosaccharides are simple sugars, like glucose. Special enzymes bind these small monomers together creating large sugar polymers, or polysaccharides. A polysaccharide is also called a glycan. A polysaccharide can be a homopolysaccharide, in which all the monosaccharides are the same, or a heteropolysaccharide in which the monosaccharides vary. Depending on which monosaccharides are connected, and which carbons in the monosaccharides connects, polysaccharides take on a variety of forms. A molecule with a straight chain of monosaccharides is called a linear polysaccharide, while a chain that has arms and turns is known as a branched polysaccharide.

Functions of Polysaccharide

Depending on their structure, polysaccharides can have a wide variety of functions in nature. Some polysaccharides are used for storing energy, some for sending cellular messages, and others for providing support to cells and tissues.

Storage of Energy

Many polysaccharides are used to store energy in organisms. While the enzymes that produce energy only work on the monosaccharides stored in a polysaccharide, polysaccharides typically fold together and can contain many monosaccharides in a dense area. Further, as the side chains of the monosaccharides form as many hydrogen bonds as possible with themselves, water cannot intrude the molecules, making them hydrophobic. This property allows the molecules to stay together and not dissolve into the cytosol. This lowers the sugar concentration in a cell, and more sugar can then be taken in. Not only do polysaccharides store the energy, but they allow for changes in the concentration gradient, which can influence cellular uptake of nutrients and water.

Cellular Communication

Many polysaccharides become glycoconjugates when they become covalently bonded to proteins or lipids. Glycolipids and glycoproteins can be used to send signals between and within cells. Proteins headed for a specific organelle may be "tagged" by certain polysaccharides that help the cell move it to a specific organelle. The polysaccharides can be identified by special proteins, which then help bind the protein, vesicle, or other substance to a microtubule. The system of microtubules and associated proteins within cells can take any substance to its destined location once tagged by specific polysaccharides. Further, multi-cellular organisms have immune systems driven by the recognition of glycoproteins on the surface of cells. The cells of a single organisms will produce specific polysaccharides to adorn its cells with. When the immune system recognizes other polysaccharides and different glycoproteins, it is set into action, and destroys the invading cells.

Cellular Support

By far one of the largest roles of polysaccharides is that of support. All plants on Earth are supported, in part, by the polysaccharide cellulose. Other organisms, like insects and fungi, use chitin to support the extracellular matrix around their cells. A polysaccharide can be mixed with any number of other components to create tissues that are more rigid, less rigid, or even materials with special properties. Between chitin and cellulose, both polysaccharides made of glucose monosaccharides, hundreds of billions of tons are created by living organisms every year. Everything from the wood in trees, to the shells of sea creatures is produced by some form of polysaccharide. Simply by rearranging the structure, polysaccharides can go from storage molecules to much stronger fibrous molecules.

Structure of a Polysaccharide

All polysaccharides are formed by the same basic process: monosaccharides are connected via glycosidic bonds. When in a polysaccharide, individual monosaccharides are known as residues.

Depending on the polysaccharide, any combination of them can be combined in series.

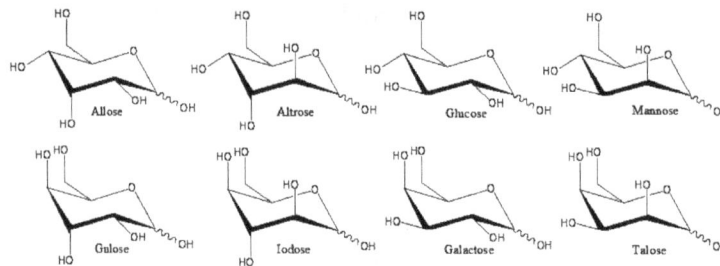

The structure of the molecules being combined determines the structures and properties of the resulting polysaccharide. The complex interaction between their hydroxyl groups (OH), other side groups, the configurations of the molecules, and the enzymes involved all affect the resulting polysaccharide produced. A polysaccharide used for energy storage will give easy access to the monosaccharides, while maintaining a compact structure. A polysaccharide used for support is usually assembled as a long chain of monosaccharides, which acts as a fiber. Many fibers together produce hydrogen bonds between fibers that strengthen the overall structure of the material.

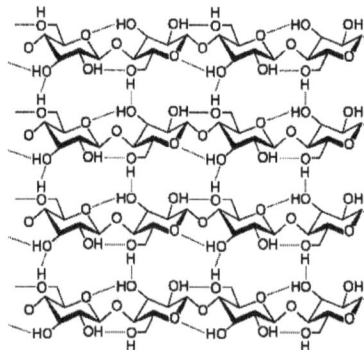

The glycosidic bonds between monosaccharides consist of an oxygen molecule bridging two carbon rings. The bond is formed when a Hydroxyl group is lost from the carbon of one molecule, while the hydrogen is lost by the hydroxyl group of another monosaccharide. The carbon on the first molecule will substitute the oxygen from the second molecule as its own, and glycosidic bond is formed. Because two molecules of hydrogen and one oxygen is expelled, the reaction produced a water molecule as well. This type of reaction is called a dehydration reaction as water is removed from the reactants.

Examples of Polysaccharides

Cellulose

Cellulose is the substance that makes up most of a plant's cell walls. Since it is made by all plants, it is probably the most abundant organic compound on Earth. Aside from being the primary building material for plants, cellulose has many others uses. According to how it is treated, cellulose can be used to make paper, film, explosives, and plastics, in addition to having many other industrial uses. For humans, cellulose is also a major source of needed fiber in our diet.

Structure of Cellulose

Cellulose is usually described by chemists and biologists as a complex carbohydrate (pronounced car-bow-HI-drayt). Carbohydrates are organic compounds made up of carbon, hydrogen, and oxygen that function as sources of energy for living things. Plants are able to make their own carbohydrates that they use for energy and to build their cell walls. According to how many atoms they have, there are several different types of carbohydrates, but the simplest and most common in a plant is glucose. Plants make glucose (formed by photosynthesis) to use for energy or to store as starch for later use. A plant uses glucose to make cellulose when it links many simple units of glucose together to form long chains. These long chains are called polysaccharides.

Scanning electron micrograph of wood cellulose. and they form very long molecules that plants use to build their walls.

It is because of these long molecules that cellulose is insoluble or does not dissolve easily in water. These long molecules also are formed into a criss-cross mesh that gives strength and shape to the cell wall. Thus while some of the food that a plant makes when it converts light energy into chemical energy (photosynthesis) is used as fuel and some is stored, the rest is turned into cellulose that serves as the main building material for a plant. Cellulose is ideal as a structural material since its fibers give strength and toughness to a plant's leaves, roots, and stems.

Cellulose and Plant Cells

Since cellulose is the main building material out of which plants are made, and plants are the primary or first link in what is known as the food chain (which describes the feeding relationships of all living things), cellulose is a very important substance.

Human uses of Cellulose

Cellulose is one of the most widely used natural substances and has become one of the most important commercial raw materials. The major sources of cellulose are plant fibers (cotton, hemp, flax, and jute are almost all cellulose) and, of course, wood (about 42 percent cellulose). Since cellulose is insoluble in water, it is easily separated from the other constituents of a plant. Cellulose has been used to make paper since the Chinese first invented the process around A.D. 100. Cellulose is separated from wood by a pulping process that grinds woodchips under flowing water. The pulp that remains is then washed, bleached, and poured over a vibrating mesh. When the water finally drains from the pulp, what remains is an interlocking web of fibers that, when dried, pressed, and smoothed, becomes a sheet of paper.

Raw cotton is 91 percent cellulose, and its fiber cells are found on the surface of the cotton seed. There are thousands of fibers on each seed, and as the cotton pod ripens and bursts open, these fiber cells die. Because these fiber cells are primarily cellulose, they can be twisted to form thread or yarn that is then woven to make cloth. Since cellulose reacts easily to both strong bases and acids, a chemical process is often used to make other products. For example, the fabric known as rayon and the transparent sheet of film called cellophane are made using a many-step process that involves an acid bath. In mixtures if nitric and sulfuric acids, cellulose can form what is called guncotton or cellulose nitrates that are used for explosives. However, when mixed with camphor, cellulose produces a plastic known as celluloid, which was used for early motion-picture film. However, because it was highly flammable (meaning it could easily catch fire), it was eventually replaced by newer and more stable plastic materials. Although cellulose is still an important natural resource, many of the products that were made from it are being produced easier and cheaper using other materials.

Starch

Starch is the most important source of carbohydrates in the human diet and accounts for more than 50% of our carbohydrate intake. It occurs in plants in the form of granules, and these are particularly abundant in seeds (especially the cereal grains) and tubers, where they serve as a storage form of carbohydrates. The breakdown of starch to glucose nourishes the plant during periods of reduced photosynthetic activity. We often think of potatoes as a "starchy" food, yet other plants contain a much greater percentage of starch (potatoes 15%, wheat 55%, corn 65%, and rice 75%). Commercial starch is a white powder.

Starch is a mixture of two polymers: amylose and amylopectin. Natural starches consist of about

10%–30% amylase and 70%–90% amylopectin. Amylose is a linear polysaccharide composed entirely of D-glucose units joined by the α-1,4-glycosidic linkages we saw in maltose. Experimental evidence indicates that amylose is not a straight chain of glucose units but instead is coiled like a spring, with six glucose monomers per turn. When coiled in this fashion, amylose has just enough room in its core to accommodate an iodine molecule. The characteristic blue-violet color that appears when starch is treated with iodine is due to the formation of the amylose-iodine complex. This color test is sensitive enough to detect even minute amounts of starch in solution.

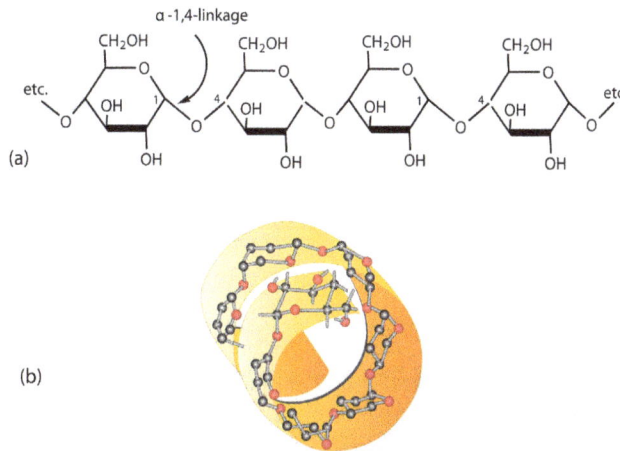

Figure: Amylose. (a) Amylose is a linear chain of α-D-glucose units joined together by α-1,4-glycosidic bonds. (b) Because of hydrogen bonding, amylose acquires a spiral structure that contains six glucose units per turn.

Amylopectin is a branched-chain polysaccharide composed of glucose units linked primarily by α-1,4-glycosidic bonds but with occasional α-1,6-glycosidic bonds, which are responsible for the branching. A molecule of amylopectin may contain many thousands of glucose units with branch points occurring about every 25–30 units. The helical structure of amylopectin is disrupted by the branching of the chain, so instead of the deep blue-violet color amylose gives with iodine, amylopectin produces a less intense reddish brown.

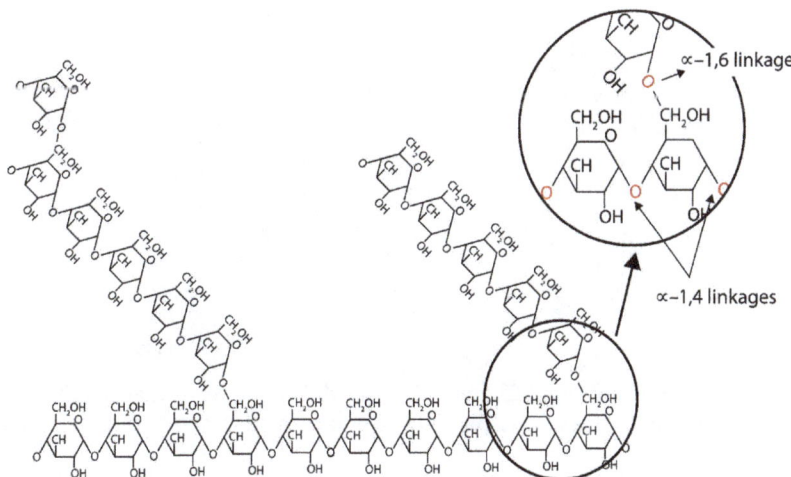

Figure: Representation of the Branching in Amylopectin and Glycogen. Both amylopectin and glycogen contain branch points that are linked through α-1,6-linkages. These branch points occur more often in glycogen.

Dextrins are glucose polysaccharides of intermediate size. The shine and stiffness imparted to clothing by starch are due to the presence of dextrins formed when clothing is ironed. Because of their characteristic stickiness with wetting, dextrins are used as adhesives on stamps, envelopes, and labels; as binders to hold pills and tablets together; and as pastes. Dextrins are more easily digested than starch and are therefore used extensively in the commercial preparation of infant foods.

The complete hydrolysis of starch yields, in successive stages, glucose:

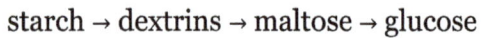

starch → dextrins → maltose → glucose

In the human body, several enzymes known collectively as amylases degrade starch sequentially into usable glucose units.

Properties

Structure

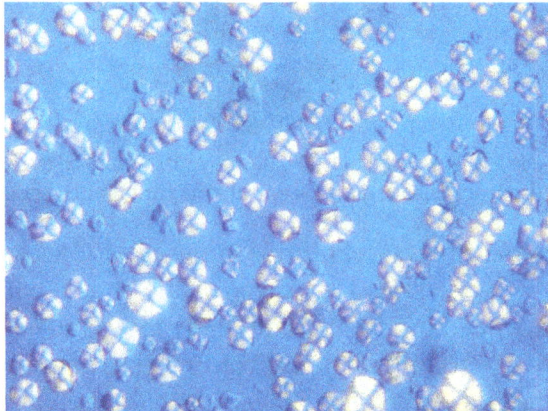

Starch, 800x magnified, under polarized light, showing characteristic extinction cross.

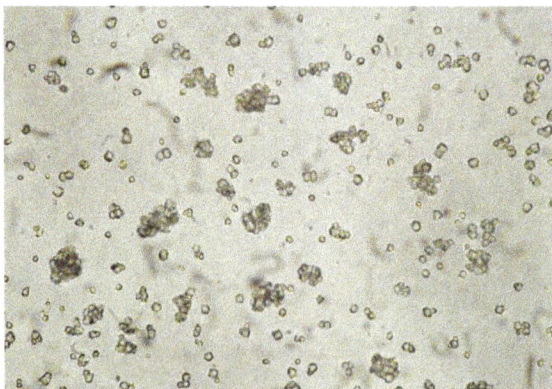

Rice starch seen on light microscope. Characteristic for the rice starch is that starch granules have an angular outline and some of them are attached to each other and form larger granules.

While amylose was thought to be completely unbranched, it is now known that some of its molecules contain a few branch points. Amylose is a much smaller molecule than amylopectin. About one quarter of the mass of starch granules in plants consist of amylose, although there are about 150 times more amylose than amylopectin molecules.

Starch molecules arrange themselves in the plant in semi-crystalline granules. Each plant species has a unique starch granular size: rice starch is relatively small (about 2 μm) while potato starches have larger granules (up to 100 μm).

Starch becomes soluble in water when heated. The granules swell and burst, the semi-crystalline structure is lost and the smaller amylose molecules start leaching out of the granule, forming a network that holds water and increasing the mixture's viscosity. This process is called starch gelatinization. During cooking, the starch becomes a paste and increases further in viscosity. During cooling or prolonged storage of the paste, the semi-crystalline structure partially recovers and the starch paste thickens, expelling water. This is mainly caused by retrogradation of the amylose. This process is responsible for the hardening of bread or staling, and for the water layer on top of a starch gel (syneresis).

Some cultivated plant varieties have pure amylopectin starch without amylose, known as *waxy starches*. The most used is waxy maize, others are glutinous rice and waxy potato starch. Waxy starches have less retrogradation, resulting in a more stable paste. High amylose starch, amylomaize, is cultivated for the use of its gel strength and for use as a resistant starch (a starch that resists digestion) in food products.

Synthetic amylose made from cellulose has a well-controlled degree of polymerization. Therefore, it can be used as a potential drug deliver carrier.

Certain starches, when mixed with water, will produce a non-newtonian fluid sometimes nicknamed "oobleck".

Hydrolysis

The enzymes that break down or hydrolyze starch into the constituent sugars are known as amylases.

Alpha-amylases are found in plants and in animals. Human saliva is rich in amylase, and the pancreas also secretes the enzyme. Individuals from populations with a high-starch diet tend to have more amylase genes than those with low-starch diets;

Beta-amylase cuts starch into maltose units. This process is important in the digestion of starch and is also used in brewing, where amylase from the skin of seed grains is responsible for converting starch to maltose (Malting, Mashing).

Given a heat of combustion of glucose of 2,805 kilojoules per mole (670 kcal/mol) whereas that of starch is 2,835 kJ (678 kcal) per mole of glucose monomer, hydrolysis releases about 30 kJ (7.2 kcal) per mole, or 166 J (40 cal) per gram of glucose product.

Dextrinization

If starch is subjected to dry heat, it breaks down to form dextrins, also called "pyrodextrins" in this context. This break down process is known as dextrinization. (Pyro)dextrins are mainly yellow to brown in color and dextrinization is partially responsible for the browning of toasted bread.

Chemical Tests

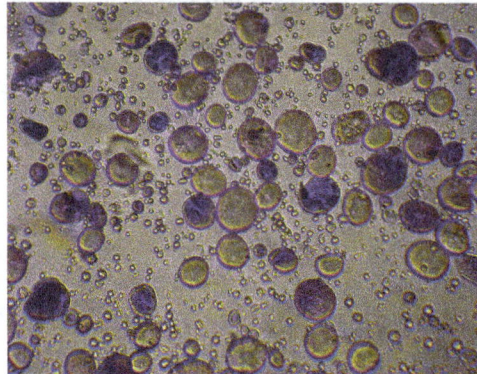

Granules of wheat starch, stained with iodine, photographed through a light microscope.

A triiodide (I_3^-) solution formed by mixing iodine and iodide (usually from potassium iodide) is used to test for starch; a dark blue color indicates the presence of starch. The details of this reaction are not fully known, but recent scientific work using single crystal x-ray crystallography and comparative Raman spectroscopy suggests that the final starch-iodine structure is similar to an infinite polyiodide chain like one found in a pyrroloperylene-iodine complex. The strength of the resulting blue color depends on the amount of amylose present. Waxy starches with little or no amylose present will color red. Benedict's test and Fehling's test is also done to indicate the presence of starch.

Starch indicator solution consisting of water, starch and iodide is often used in redox titrations: in the presence of an oxidizing agent the solution turns blue, in the presence of reducing agent the blue color disappears because triiodide (I_3^-) ions break up into three iodide ions, disassembling the starch-iodine complex. Starch solution was used as indicator for visualizing the periodic formation and consumption of triiodide intermediate in the Briggs-Rauscher oscillating reaction. The starch, however, changes the kinetics of the reaction steps involving triiodide ion. A 0.3% w/w solution is the standard concentration for a starch indicator. It is made by adding 3 grams of soluble starch to 1 liter of heated water; the solution is cooled before use (starch-iodine complex becomes unstable at temperatures above 35 °C).

Each species of plant has a unique type of starch granules in granular size, shape and crystallization pattern. Under the microscope, starch grains stained with iodine illuminated from behind with polarized light show a distinctive Maltese cross effect (also known as extinction cross and birefringence).

Food

Starch is the most common carbohydrate in the human diet and is contained in many staple foods. The major sources of starch intake worldwide are the cereals (rice, wheat, and maize) and the root vegetables (potatoes and cassava). Many other starchy foods are grown, some only in specific climates, including acorns, arrowroot, arracacha, bananas, barley, breadfruit, buckwheat, canna, colocasia, katakuri, kudzu, malanga, millet, oats, oca, polynesian arrowroot, sago, sorghum, sweet potatoes, rye, taro, chestnuts, water chestnuts and yams, and many kinds of beans, such as favas, lentils, mung beans, peas, and chickpeas.

Widely used prepared foods containing starch are bread, pancakes, cereals, noodles, pasta, porridge and tortilla.

Digestive enzymes have problems digesting crystalline structures. Raw starch is digested poorly in the duodenum and small intestine, while bacterial degradation takes place mainly in the colon. When starch is cooked, the digestibility is increased.

Starch gelatinization during cake baking can be impaired by sugar competing for water, preventing gelatinization and improving texture.

Before the advent of processed foods, people consumed large amounts of uncooked and unprocessed starch-containing plants, which contained high amounts of resistant starch. Microbes within the large intestine fermented the starch, produced short-chain fatty acids, which are used as energy, and support the maintenance and growth of the microbes. More highly processed foods are more easily digested and release more glucose in the small intestine—less starch reaches the large intestine and more energy is absorbed by the body. It is thought that this shift in energy delivery (as a result of eating more processed foods) may be one of the contributing factors to the development of metabolic disorders of modern life, including obesity and diabetes.

Starch Production

The starch industry extracts and refines starches from seeds, roots and tubers, by wet grinding, washing, sieving and drying. Today, the main commercial refined starches are cornstarch, tapioca, arrowroot, and wheat, rice, and potato starches. To a lesser extent, sources of refined starch are sweet potato, sago and mung bean. To this day, starch is extracted from more than 50 types of plants.

Untreated starch requires heat to thicken or gelatinize. When a starch is pre-cooked, it can then be used to thicken instantly in cold water. This is referred to as a pregelatinized starch.

Starch Sugars

Starch can be hydrolyzed into simpler carbohydrates by acids, various enzymes, or a combination of the two. The resulting fragments are known as dextrins. The extent of conversion is typically quantified by *dextrose equivalent* (DE), which is roughly the fraction of the glycosidic bonds in starch that have been broken.

These starch sugars are by far the most common starch based food ingredient and are used as sweeteners in many drinks and foods. They include:

- Maltodextrin, a lightly hydrolyzed (DE 10–20) starch product used as a bland-tasting filler and thickener.

- Various glucose syrups (DE 30–70), also called corn syrups in the US, viscous solutions used as sweeteners and thickeners in many kinds of processed foods.

- Dextrose (DE 100), commercial glucose, prepared by the complete hydrolysis of starch.

- High fructose syrup, made by treating dextrose solutions with the enzyme glucose isomerase, until a substantial fraction of the glucose has been converted to fructose. In the United

States sugar prices are two to three times higher than in the rest of the world; high-fructose corn syrup is significantly cheaper, and is the principal sweetener used in processed foods and beverages. Fructose also has better microbiological stability. One kind of high fructose corn syrup, HFCS-55, is sweeter than sucrose because it is made with more fructose, while the sweetness of HFCS-42 is on par with sucrose.

- Sugar alcohols, such as maltitol, erythritol, sorbitol, mannitol and hydrogenated starch hydrolysate, are sweeteners made by reducing sugars.

Modified Starches

A modified starch is a starch that has been chemically modified to allow the starch to function properly under conditions frequently encountered during processing or storage, such as high heat, high shear, low pH, freeze/thaw and cooling.

The modified food starches are E coded according to the International Numbering System for Food Additives (INS):

- 1400 Dextrin
- 1401 Acid-treated starch
- 1402 Alkaline-treated starch
- 1403 Bleached starch
- 1404 Oxidized starch
- 1405 Starches, enzyme-treated
- 1410 Monostarch phosphate
- 1412 Distarch phosphate
- 1413 Phosphated distarch phosphate
- 1414 Acetylated distarch phosphate
- 1420 Starch acetate
- 1422 Acetylated distarch adipate
- 1440 Hydroxypropyl starch
- 1442 Hydroxypropyl distarch phosphate
- 1443 Hydroxypropyl distarch glycerol
- 1450 Starch sodium octenyl succinate
- 1451 Acetylated oxidized starch

INS 1400, 1401, 1402, 1403 and 1405 are in the EU food ingredients without an E-number. Typical modified starches for technical applications are cationic starches, hydroxyethyl starch and carboxymethylated starches.

Use as Food Additive

As an additive for food processing, food starches are typically used as thickeners and stabilizers in foods such as puddings, custards, soups, sauces, gravies, pie fillings, and salad dressings, and to make noodles and pastas. Function as thickeners, extenders, emulsion stabilizers and are exceptional binders in processed meats.

Gummed sweets such as jelly beans and wine gums are not manufactured using a mold in the conventional sense. A tray is filled with native starch and leveled. A positive mold is then pressed into the starch leaving an impression of 1,000 or so jelly beans. The jelly mix is then poured into the impressions and put onto a stove to set. This method greatly reduces the number of molds that must be manufactured.

Use in Pharmaceutical Industry

In the pharmaceutical industry, starch is also used as an excipient, as tablet disintegrant, and as binder.

Resistant Starch

Resistant starch is starch that escapes digestion in the small intestine of healthy individuals. High amylose starch from corn has a higher gelatinization temperature than other types of starch and retains its resistant starch content through baking, mild extrusion and other food processing techniques. It is used as an insoluble dietary fiber in processed foods such as bread, pasta, cookies, crackers, pretzels and other low moisture foods. It is also utilized as a dietary supplement for its health benefits. Published studies have shown that resistant starch helps to improve insulin sensitivity, increases satiety and improves markers of colonic function. It has been suggested that resistant starch contributes to the health benefits of intact whole grains.

Industrial Applications

Starch adhesive.

Papermaking

Papermaking is the largest non-food application for starches globally, consuming many millions of metric tons annually. In a typical sheet of copy paper for instance, the starch content may be as high as 8%. Both chemically modified and unmodified starches are used in papermaking. In the wet part of the papermaking process, generally called the "wet-end", the starches used are cationic and have a positive charge bound to the starch polymer. These starch derivatives associate with the anionic or negatively charged paper fibers/cellulose and inorganic fillers. Cationic starches together with other retention and internal sizing agents help to give the necessary strength properties to the paper web formed in the papermaking process (wet strength), and to provide strength to the final paper sheet (dry strength).

In the dry end of the papermaking process, the paper web is rewetted with a starch based solution. The process is called surface sizing. Starches used have been chemically, or enzymatically depolymerized at the paper mill or by the starch industry (oxidized starch). The size/starch solutions are applied to the paper web by means of various mechanical presses (size presses). Together with surface sizing agents the surface starches impart additional strength to the paper web and additionally provide water hold out or "size" for superior printing properties. Starch is also used in paper coatings as one of the binders for the coating formulations which include a mixture of pigments, binders and thickeners. Coated paper has improved smoothness, hardness, whiteness and gloss and thus improves printing characteristics.

Corrugated Board Adhesives

Corrugated board adhesives are the next largest application of non-food starches globally. Starch glues are mostly based on unmodified native starches, plus some additive such as borax and caustic soda. Part of the starch is gelatinized to carry the slurry of uncooked starches and prevent sedimentation. This opaque glue is called a SteinHall adhesive. The glue is applied on tips of the fluting. The fluted paper is pressed to paper called liner. This is then dried under high heat, which causes the rest of the uncooked starch in glue to swell/gelatinize. This gelatinizing makes the glue a fast and strong adhesive for corrugated board production.

Clothing Starch

Clothing or laundry starch is a liquid prepared by mixing a vegetable starch in water (earlier preparations also had to be boiled), and is used in the laundering of clothes. Starch was widely used in Europe in the 16th and 17th centuries to stiffen the wide collars and ruffs of fine linen which surrounded the necks of the well-to-do. During the 19th and early 20th century it was stylish to stiffen the collars and sleeves of men's shirts and the ruffles of women's petticoats by applying starch to them as the clean clothes were being ironed. Starch gave clothing smooth, crisp edges, and had an additional practical purpose: dirt and sweat from a person's neck and wrists would stick to the starch rather than to the fibers of the clothing. The dirt would wash away along with the starch; after laundering, the starch would be reapplied. Today, starch is sold in aerosol cans for home use.

Other

Another large non-food starch application is in the construction industry, where starch is used in the gypsum wall board manufacturing process. Chemically modified or unmodified starches are added to the stucco containing primarily gypsum. Top and bottom heavyweight sheets of paper are applied to the formulation, and the process is allowed to heat and cure to form the eventual rigid wall board. The starches act as a glue for the cured gypsum rock with the paper covering, and also provide rigidity to the board.

Starch is used in the manufacture of various adhesives or glues for book-binding, wallpaper adhesives, paper sack production, tube winding, gummed paper, envelope adhesives, school glues and bottle labeling. Starch derivatives, such as yellow dextrins, can be modified by addition of some chemicals to form a hard glue for paper work; some of those forms use borax or soda ash, which are mixed with the starch solution at 50–70 °C (122–158 °F) to create a very good adhesive. Sodium silicate can be added to reinforce these formula.

- Textile chemicals from starch: warp sizing agents are used to reduce breaking of yarns during weaving. Starch is mainly used to size cotton based yarns. Modified starch is also used as textile printing thickener.

- In oil exploration, starch is used to adjust the viscosity of drilling fluid, which is used to lubricate the drill head and suspend the grinding residue in petroleum extraction.

- Starch is also used to make some packing peanuts, and some drop ceiling tiles.

- In the printing industry, food grade starch is used in the manufacture of anti-set-off spray powder used to separate printed sheets of paper to avoid wet ink being set off.

- For body powder, powdered corn starch is used as a substitute for talcum powder, and similarly in other health and beauty products.

- Starch is used to produce various bioplastics, synthetic polymers that are biodegradable. An example is polylactic acid based on glucose from starch.

- Glucose from starch can be further fermented to biofuel corn ethanol using the so-called wet milling process. Today most bioethanol production plants use the dry milling process to ferment corn or other feedstock directly to ethanol.

- Hydrogen production could use glucose from starch as the raw material, using enzymes.

Glycogen

Glycogen is a large, branched polysaccharide that is the main storage form of glucose in animals and humans. Glycogen is as an important energy reservoir; when energy is required by the body, glycogen in broken down to glucose, which then enters the glycolytic or pentose phosphate pathway or is released into the bloodstream. Glycogen is also an important form of glucose storage in fungi and bacteria.

Glycogen Structure

Glycogen is a branched polymer of glucose. Glucose residues are linked linearly by α-1,4 glycosidic bonds, and approximately every ten residues a chain of glucose residues branches off via α-1,6 glycosidic linkages. The α-glycosidic bonds give rise to a helical polymer structure. Glycogen is hydrated with three to four parts water and forms granules in the cytoplasm that are 10-40nm in diameter. The protein glycogenin, which is involved in glycogen synthesis, is located at the core of each glycogen granule. Glycogen is an analogue of starch, which is the main form of glucose storage in most plants, but starch has fewer branches and is less compact than glycogen.

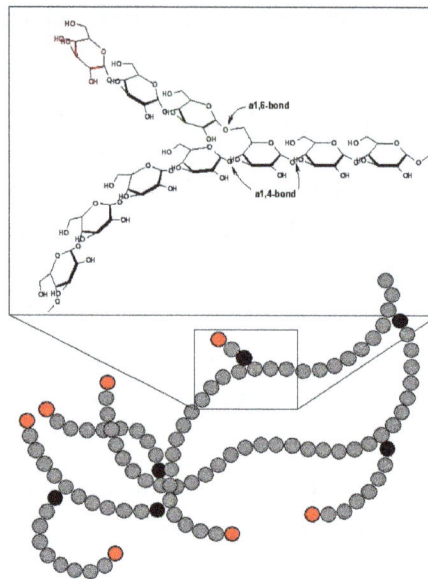

Structure of glycogen: Green circles represent α-1,6 linkages at branch points, and red circles represent the nonreducing ends of the chain.

Glycogen Function

In animals and humans, glycogen is found mainly in muscle and liver cells. Glycogen is synthesized from glucose when blood glucose levels are high, and serves as a ready source of glucose for tissues throughout the body when blood glucose levels decline.

Liver Cells

Glycogen makes up 6-10% of the liver by weight. When food is ingested, blood glucose levels rise, and insulin released from the pancreas promotes the uptake of glucose into liver cells. Insulin also activates enzymes involved in glycogen synthesis, such as glycogen synthase. While glucose and insulin levels are sufficiently high, glycogen chains are elongated by the addition of glucose molecules, a process termed glyconeogenesis. As glucose and insulin levels decrease, glycogen synthesis ceases. When blood glucose levels fall below a certain level, glucagon released from the pancreas signals to liver cells to break down glycogen. Glycogen is broken down via glycogenolysis into glucose-1-phosphate, which is converted to glucose and released into the bloodstream. Thus, glycogen serves as the main buffer of blood glucose levels by storing glucose when it levels are high and releasing glucose when levels are low. Glycogen breakdown in the liver is critical for supplying glucose to meet the body's energetic needs. In addition to glucagon, cortisol, epinephrine, and norepinephrine also stimulate glycogen breakdown.

Muscle Cells

In contrast to liver cells, glycogen only accounts for 1-2% of muscle by weight. However, given the greater mass of muscle in the body, the total amount of glycogen stored in muscle is greater than that stored in liver. Muscle also differs from liver in that the glycogen in muscle only provides glucose to the muscle cell itself. Muscle cells do not express the enzyme glucose-6-phosphatase, which is required to release glucose into the bloodstream. The glucose-1-phosphate produced from glycogen breakdown in muscle fibers is converted to glucose-6-phosphate and provides energy to the muscle during a bout of exercise or in response to stress, as in the fight-or-flight response.

Other Tissues

In addition to liver and muscle, glycogen in found in smaller amounts in other tissues, including red blood cells, white blood cells, kidney cells, and some glial cells. Additionally, glycogen is used to store glucose in the uterus to provide for the energetic needs of the embryo.

Fungi and Bacteria

Microorganisms possess mechanisms for storing energy to cope in the event of limited environmental resources, and glycogen represents a main energy storage form. Nutrient limitation (low levels of carbon, phosphorus, nitrogen, or sulfur) can stimulate glycogen formation in yeast, while bacteria synthesize glycogen in response to readily available carbon energy sources with limitation of other nutrients. Bacterial growth and yeast sporulation have also been associated with glycogen accumulation.

Glycogen Metabolism

Glycogen homeostasis is a highly regulated process that allows the body to store or release glucose depending on its energetic needs. The basic steps in glucose metabolism are glycogenesis, or glycogen synthesis, and glycogenolysis, or glycogen breakdown.

Glycogenesis

Glycogen synthesis requires energy, which is supplied by uridine triphosphate (UTP). Hexokinases or glucokinase first phosphorylate free glucose to form glucose-6-phosphate, which is converted to glucose-1-phosphate by phosphoglucomutase. UTP-glucose-1-phosphate uridylyltransferase then catalyzes the activation of glucose, in which UTP and glucose-1-phosphate react to form UDP-glucose. In de novo glycogen synthesis, the protein glycogenin catalyzes the attachment of UDP-glucose to itself. Glycogenin is a homodimer containing a tyrosine residue in each subunit that serves as an anchor or attachment point for glucose. Additional glucose molecules are subsequently added to the reducing end of the previous glucose molecule to form a chain of approximately eight glucose molecules. Glycogen synthase then extends the chain by adding glucose via α-1,4 glycosidic linkages.

Branching is catalyzed by amylo-(1,4 to 1,6)-transglucosidase, also called the glycogen branching enzyme. The glycogen branching enzyme transfers a fragment of six to seven glucose molecules from the end of a chain to the C6 of a glucose molecule located further inside the glycogen molecule, forming α-1,6 glycosidic linkages.

Glycogenolysis

Glucose is removed from glycogen by glycogen phosphorylase, which phosphorolytically removes one molecule of glucose from the nonreducing end, yielding glucose-1-phosphate. The glucose-1-phosphate generated by glycogen breakdown is converted to glucose-6-phosphate, a process that requires the enzyme phosphoglucomutase. Phosphoglucomutase transfers a phosphate group from a phosphorylated serine residue within the active site to C6 of glucose-1-phosphate, producing glucose-1,6-bisphosphate. The glucose C1 phosphate is then attached to the active site serine within phosphoglucomutase, and glucose-6-phosphate is released.

Glycogen phosphorylase is not able to cleave glucose from branch points; debranching requires amylo-1,6-glucosidase, 4-α-glucanotransferase, or glycogen debranching enzyme (GDE), which has glucotransferase and glucosidase activities. About four residues from a branch point, glycogen phosphorylase is unable to remove glucose residues. GDE cleaves the final three residues of a branch and attaches them to C4 of a glucose molecule at the end of a different branch, then removes the final α-1,6-linked glucose residue from the branch point. GDE does not remove the α-1,6-linked glucose from the branch point phosphorylytically, meaning that free glucose is released. This free glucose could in theory be released from muscle into the bloodstream without the action of glucose-6-phosphatase; however this free glucose is rapidly phosphorylated by hexokinase, preventing it from entering the bloodstream.

The glucose-6-phosphate resulting from glycogen breakdown may be converted to glucose by the action of glucose-6-phosphatase and released into the bloodstream. This occurs in liver, intestine, and kidney, but not in muscle, where this enzyme is absent. In muscle, glucose-6-phosphate enters the glycolytic pathway and provides energy to the cell. Glucose-6-phosphate may also enter the pentose phosphate pathway, resulting in the production of NADPH and five carbon sugars.

Carbohydrate Metabolism

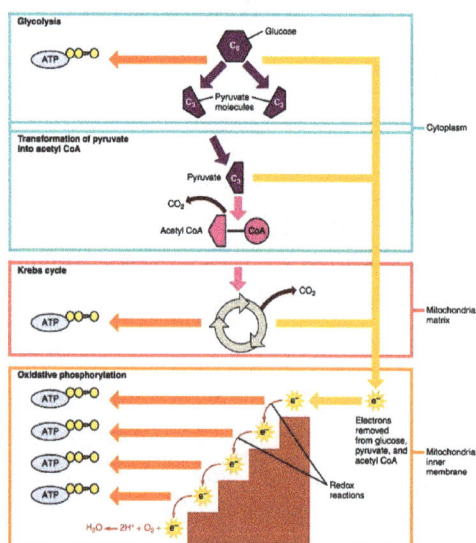

Figure: Cellular respiration oxidizes glucose molecules through glycolysis, the Krebs cycle, and oxidative phosphorylation to produce ATP.

Carbohydrates are organic molecules composed of carbon, hydrogen, and oxygen atoms. The family of carbohydrates includes both simple and complex sugars. Glucose and fructose are examples of simple sugars, and starch, glycogen, and cellulose are all examples of complex sugars. The complex sugars are also called polysaccharides and are made of multiple monosaccharide molecules. Polysaccharides serve as energy storage (e.g., starch and glycogen) and as structural components (e.g., chitin in insects and cellulose in plants).

During digestion, carbohydrates are broken down into simple, soluble sugars that can be transported across the intestinal wall into the circulatory system to be transported throughout the body. Carbohydrate digestion begins in the mouth with the action of salivary amylase on starches and ends with monosaccharides being absorbed across the epithelium of the small intestine. Once the absorbed monosaccharides are transported to the tissues, the process of cellular respiration begins.

Glycolysis

Glucose is the body's most readily available source of energy. After digestive processes break polysaccharides down into monosaccharides, including glucose, the monosaccharides are transported across the wall of the small intestine and into the circulatory system, which transports them to the liver. In the liver, hepatocytes either pass the glucose on through the circulatory system or store excess glucose as glycogen. Cells in the body take up the circulating glucose in response to insulin and, through a series of reactions called glycolysis, transfer some of the energy in glucose to ADP to form ATP. The last step in glycolysis produces the product pyruvate.

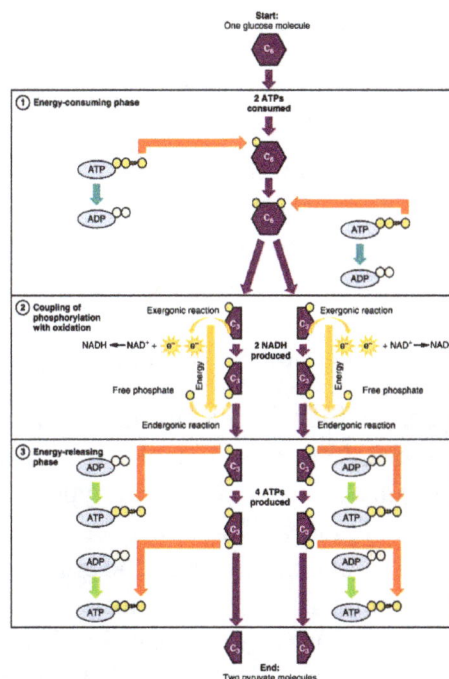

Figure: During the energy-consuming phase of glycolysis, two ATPs are consumed, transferring two phosphates to the glucose molecule.

Glycolysis begins with the phosphorylation of glucose by hexokinase to form glucose-6-phosphate. This step uses one ATP, which is the donor of the phosphate group. Under the action of phosphofructokinase, glucose-6-phosphate is converted into fructose-6-phosphate. At this point, a second ATP donates its phosphate group, forming fructose-1,6-bisphosphate. This six-carbon sugar is split to form two phosphorylated three-carbon molecules, glyceraldehyde-3-phosphate and dihydroxyacetone phosphate, which are both converted into glyceraldehyde-3-phosphate. The glyceraldehyde-3-phosphate is further phosphorylated with groups donated by dihydrogen phosphate present in the cell to form the three-carbon molecule 1,3-bisphosphoglycerate. The energy of this reaction comes from the oxidation of (removal of electrons from) glyceraldehyde-3-phosphate. In a series of reactions leading to pyruvate, the two phosphate groups are then transferred to two ADPs to form two ATPs. Thus, glycolysis uses two ATPs but generates four ATPs, yielding a net gain of two ATPs and two molecules of pyruvate. In the presence of oxygen, pyruvate continues on to the Krebs cycle (also called the citric acid cycle or tricarboxylic acid cycle (TCA), where additional energy is extracted and passed on.

The glucose molecule then splits into two three-carbon compounds, each containing a phosphate. During the second phase, an additional phosphate is added to each of the three-carbon compounds. The energy for this endergonic reaction is provided by the removal (oxidation) of two electrons from each three-carbon compound. During the energy-releasing phase, the phosphates are removed from both three-carbon compounds and used to produce four ATP molecules.

Glycolysis can be divided into two phases: energy consuming (also called chemical priming) and energy yielding. The first phase is the energy-consuming phase, so it requires two ATP molecules to start the reaction for each molecule of glucose. However, the end of the reaction produces four ATPs, resulting in a net gain of two ATP energy molecules.

Glycolysis can be expressed as the following equation:

$$\text{Glucose} + 2\,\text{ATP} + 2\,\text{NAD}^+ + 4\,\text{ADP} + 2\,P_i \rightarrow \text{Pyruvate} + 4\,\text{ATP} + 2\,\text{NADH} + 2\,\text{H}^+$$

This equation states that glucose, in combination with ATP (the energy source), NAD+ (a coenzyme that serves as an electron acceptor), and inorganic phosphate, breaks down into two pyruvate molecules, generating four ATP molecules—for a net yield of two ATP—and two energy-containing NADH coenzymes. The NADH that is produced in this process will be used later to produce ATP in the mitochondria. Importantly, by the end of this process, one glucose molecule generates two pyruvate molecules, two high-energy ATP molecules, and two electron-carrying NADH molecules.

The following discussions of glycolysis include the enzymes responsible for the reactions. When glucose enters a cell, the enzyme hexokinase (or glucokinase, in the liver) rapidly adds a phosphate to convert it into glucose-6-phosphate. A kinase is a type of enzyme that adds a phosphate molecule to a substrate (in this case, glucose, but it can be true of other molecules also). This conversion step requires one ATP and essentially traps the glucose in the cell, preventing it from passing back through the plasma membrane, thus allowing glycolysis to proceed. It also functions to maintain a concentration gradient with higher glucose levels in the blood than in the tissues. By establishing this concentration gradient, the glucose in the blood will be able to flow from an area of high concentration (the blood) into an area of low concentration (the tissues) to be either used or stored.

Hexokinase is found in nearly every tissue in the body. Glucokinase, on the other hand, is expressed in tissues that are active when blood glucose levels are high, such as the liver. Hexokinase has a higher affinity for glucose than glucokinase and therefore is able to convert glucose at a faster rate than glucokinase. This is important when levels of glucose are very low in the body, as it allows glucose to travel preferentially to those tissues that require it more.

In the next step of the first phase of glycolysis, the enzyme glucose-6-phosphate isomerase converts glucose-6-phosphate into fructose-6-phosphate. Like glucose, fructose is also a six carbon-containing sugar. The enzyme phosphofructokinase-1 then adds one more phosphate to convert fructose-6-phosphate into fructose-1-6-bisphosphate, another six-carbon sugar, using another ATP molecule. Aldolase then breaks down this fructose-1-6-bisphosphate into two three-carbon molecules, glyceraldehyde-3-phosphate and dihydroxyacetone phosphate. The triosephosphate isomerase enzyme then converts dihydroxyacetone phosphate into a second glyceraldehyde-3-phosphate molecule. Therefore, by the end of this chemical- priming or energy-consuming phase, one glucose molecule is broken down into two glyceraldehyde-3-phosphate molecules.

The second phase of glycolysis, the energy-yielding phase, creates the energy that is the product of glycolysis. Glyceraldehyde-3-phosphate dehydrogenase converts each three-carbon glyceraldehyde-3-phosphate produced during the energy-consuming phase into 1,3-bisphosphoglycerate. This reaction releases an electron that is then picked up by NAD+ to create an NADH molecule. NADH is a high-energy molecule, like ATP, but unlike ATP, it is not used as energy currency by the cell. Because there are two glyceraldehyde-3-phosphate molecules, two NADH molecules are synthesized during this step. Each 1,3-bisphosphoglycerate is subsequently dephosphorylated (i.e., a phosphate is removed) by phosphoglycerate kinase into 3-phosphoglycerate. Each phosphate released in this reaction can convert one molecule of ADP into one high- energy ATP molecule, resulting in a gain of two ATP molecules.

The enzyme phosphoglycerate mutase then converts the 3-phosphoglycerate molecules into 2-phosphoglycerate. The enolase enzyme then acts upon the 2-phosphoglycerate molecules to convert them into phosphoenolpyruvate molecules. The last step of glycolysis involves the dephosphorylation of the two phosphoenolpyruvate molecules by pyruvate kinase to create two pyruvate molecules and two ATP molecules.

In summary, one glucose molecule breaks down into two pyruvate molecules, and creates two net ATP molecules and two NADH molecules by glycolysis. Therefore, glycolysis generates energy for the cell and creates pyruvate molecules that can be processed further through the aerobic Krebs cycle (also called the citric acid cycle or tricarboxylic acid cycle); converted into lactic acid or alcohol (in yeast) by fermentation; or used later for the synthesis of glucose through gluconeogenesis.

Anaerobic Respiration

When oxygen is limited or absent, pyruvate enters an anaerobic pathway. In these reactions, pyruvate can be converted into lactic acid. In addition to generating an additional ATP, this pathway serves to keep the pyruvate concentration low so glycolysis continues, and it oxidizes NADH into the NAD+ needed by glycolysis. In this reaction, lactic acid replaces oxygen as the final electron acceptor. Anaerobic respiration occurs in most cells of the body when oxygen is limited or mitochondria are absent or nonfunctional. For example, because erythrocytes (red blood cells) lack mitochondria, they must produce their ATP from anaerobic respiration. This is an effective pathway of ATP production for short periods of time, ranging from seconds to a few minutes. The lactic acid produced diffuses into the plasma and is carried to the liver, where it is converted back into pyruvate or glucose via the Cori cycle. Similarly, when a person exercises, muscles use ATP faster than oxygen can be delivered to them. They depend on glycolysis and lactic acid production for rapid ATP production.

Aerobic Respiration

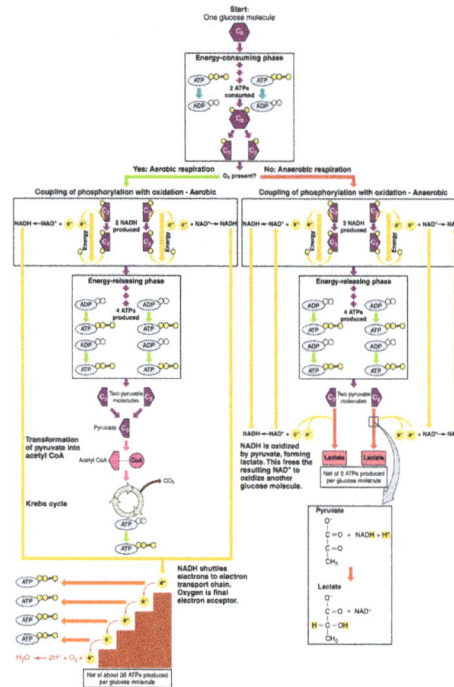

Figure: The process of anaerobic respiration converts glucose into two lactate molecules in the absence of oxygen or within erythrocytes that lack mitochondria. During aerobic respiration, glucose is oxidized into two pyruvate molecules.

In the presence of oxygen, pyruvate can enter the Krebs cycle where additional energy is extracted as electrons are transferred from the pyruvate to the receptors NAD+, GDP, and FAD, with carbon dioxide being a "waste product" The NADH and FADH2 pass electrons on to the electron transport chain, which uses the transferred energy to produce ATP. As the terminal step in the electron transport chain, oxygen is the terminal electron acceptor and creates water inside the mitochondria.

Gluconeogenesis

Gluconeogenesis is the synthesis of new glucose molecules from pyruvate, lactate, glycerol, or the amino acids alanine or glutamine. This process takes place primarily in the liver during periods of low glucose, that is, under conditions of fasting, starvation, and low carbohydrate diets. Certain key organs, including the brain, can use only glucose as an energy source; therefore, it is essential that the body maintain a minimum blood glucose concentration. When the blood glucose concentration falls below that certain point, new glucose is synthesized by the liver to raise the blood concentration to normal.

Gluconeogenesis is not simply the reverse of glycolysis. There are some important differences. Pyruvate is a common starting material for gluconeogenesis. First, the pyruvate is converted into oxaloacetate. Oxaloacetate then serves as a substrate for the enzyme phosphoenolpyruvate carboxykinase (PEPCK), which transforms oxaloacetate into phosphoenolpyruvate (PEP). From this step, gluconeogenesis is nearly the reverse of glycolysis. PEP is converted back into 2-phosphoglycerate, which is converted into 3-phosphoglycerate. Then, 3-phosphoglycerate is converted into 1,3 bisphosphoglycerate and then into glyceraldehyde-3-phosphate. Two molecules of glyceraldehyde-3-phosphate then combine to form fructose-1-6-bisphosphate, which is converted into fructose 6-phosphate and then into glucose-6-phosphate. Finally, a series of reactions generates glucose itself. In gluconeogenesis (as compared to glycolysis), the enzyme hexokinase is replaced by glucose-6-phosphatase, and the enzyme phosphofructokinase-1 is replaced by fructose-1,6-bisphosphatase. This helps the cell to regulate glycolysis and gluconeogenesis independently of each other.

References

- Configuration, carbohydrate, science: britannica.com, Retrieved April 16, 2019

- Monosaccharide, science : britannica.com , Retrieved July 19, 2019

- Monosaccharides_Structures, Monosaccharaides_and_Disaccharides, CARBOHYDRATES, Biochemistry: libretexts.org, Retrieved January 23, 2019

- Function-monosaccharide-biology: seattlepi.com, Retrieved June 16, 2019

- Glucose, entry : newworldencyclopedia.org, Retrieved May 29, 2019

- Fructose, entry : newworldencyclopedia.org, Retrieved August 21, 2019

- Galactose, carbohydrates : tuscany-diet.net, Retrieved April 12, 2019

- Disaccharide : biologydictionary.net, Retrieved June 14, 2019

- list-of-disaccharide-examples: thoughtco.com, Retrieved March 18, 2019

- Disaccharides, carbs: nutrientsreview.com, Retrieved February 26, 2019

- Sucrose: biologydictionary.net, Retrieved May 25, 2019

- Lactose: scienceofcooking.com, Retrieved July 31, 2019

- Maltose, entry: newworldencyclopedia.org, Retrieved May 3, 2019

- Polysaccharide: biologydictionary.net , Retrieved June 9, 2019

- Cellulose, Ca-Ch: scienceclarified.com , Retrieved March 13, 2019

- Starch_and_Cellulose, Stereochemistry, Organic_Chemistry: libretexts.org , Retrieved February 11, 2019

- Glycogen: biologydictionary.net, Retrieved May 17, 2019

- Carbohydrate-metabolism, suny-ap2: .lumenlearning.com , Retrieved April 20, 2019

Chapter 3

Lipid Metabolism

A biomolecule that is soluble in nonpolar solvents is known as lipid. The various categories of lipids include glycerophospholipids, fatty acids, sterol lipids, polyketides, prenol lipids, glycerolipids, sphingolipids and saccharolipids. The synthesis and degradation of lipids in cell is known as lipid metabolism. The chapter closely examines the diverse aspects of lipids and the key concepts of lipid metabolism to provide an extensive understanding of the subject.

Lipid

Lipid is any of a diverse group of organic compounds including fats, oils, hormones, and certain components of membranes that are grouped together because they do not interact appreciably with water. One type of lipid, the triglycerides, is sequestered as fat in adipose cells, which serve as the energy-storage depot for organisms and also provide thermal insulation. Some lipids such as steroid hormones serve as chemical messengers between cells, tissues, and organs, and others communicate signals between biochemical systems within a single cell. The membranes of cells and organelles (structures within cells) are microscopically thin structures formed from two layers of phospholipid molecules. Membranes function to separate individual cells from their environments and to compartmentalize the cell interior into structures that carry out special functions. So important is this compartmentalizing function that membranes, and the lipids that form them, must have been essential to the origin of life itself.

Water is the biological milieu—the substance that makes life possible—and almost all the molecular components of living cells, whether they be found in animals, plants, or microorganisms, are soluble in water. Molecules such as proteins, nucleic acids, and carbohydrates have an affinity for water and are called hydrophilic ("water-loving"). Lipids, however, are hydrophobic ("water-fearing"). Some lipids are amphipathic—part of their structure is hydrophilic and another part, usually a larger section, is hydrophobic. Amphipathic lipids exhibit a unique behaviour in water: they spontaneously form ordered molecular aggregates, with their hydrophilic ends on the outside, in contact with the water, and their hydrophobic parts on the inside, shielded from the water. This property is key to their role as the fundamental components of cellular and organelle membranes.

Although biological lipids are not large macromolecular polymers (e.g., proteins, nucleic acids, and polysaccharides), many are formed by the chemical linking of several small constituent molecules. Many of these molecular building blocks are similar, or homologous, in structure. The homologies allow lipids to be classified into a few major groups: fatty acids, fatty acid derivatives, cholesterol and its derivatives, and lipoproteins.

Biological Functions of Lipids

The majority of lipids in biological systems function either as a source of stored metabolic energy or as structural matrices and permeability barriers in biological membranes. Very small amounts of special lipids act as both intracellular messengers and extracellular messengers such as hormones and pheromones. Amphipathic lipids, the molecules that allow membranes to form compartments, must have been among the progenitors of living beings. This theory is supported by studies of several simple, single-cell organisms, in which up to one-third of the genome is thought to code for membrane proteins and the enzymes of membrane lipid biosynthesis.

Cellular Energy Source

Fatty acids that are stored in adipose tissue as triglycerides are a major energy source in higher animals, as is glucose, a simple six-carbon carbohydrate. In healthy, well-fed humans only about 2 percent of the energy is derived from the metabolism of protein. Large amounts of lipids are stored in adipose tissue. In the average American male about 25 percent of body weight is fat, whereas only 1 percent is accounted for by glycogen (a polymer of glucose). In addition, the energy available to the body from oxidative metabolism of 1 gram of triglyceride is more than twice that produced by the oxidation of an equal weight of carbohydrate such as glycogen.

Storage of Triglyceride in Adipose Cells

In higher animals and humans, adipose tissue consisting of adipocytes (fat cells) is widely distributed over the body—mainly under the skin, around deep blood vessels, and in the abdominal cavity and to a lesser degree in association with muscles. Bony fishes have adipose tissue mainly distributed among muscle fibres, but sharks and other cartilaginous fishes store lipids in the liver. The fat stored in adipose tissue arises from the dietary intake of fat or carbohydrate in excess of the energy requirements of the body. A dietary excess of 1 gram of triglyceride is stored as 1 gram of fat, but only about 0.3 gram of dietary excess carbohydrate can be stored as triglyceride. The reverse process, the conversion of excess fat to carbohydrate, is metabolically impossible. In humans, excessive dietary intake can make adipose tissue the largest mass in the body.

Excess triglyceride is delivered to the adipose tissue by lipoproteins in the blood. There the triglycerides are hydrolyzed to free fatty acids and glycerol through the action of the enzyme lipoprotein lipase, which is bound to the external surface of adipose cells. Apoprotein C-II activates this enzyme, as do the quantities of insulin that circulate in the blood following ingestion of food. The liberated free fatty acids are then taken up by the adipose cells and resynthesized into triglycerides, which accumulate in a fat droplet in each cell.

Mobilization of Fatty Acids

In times of stress when the body requires energy, fatty acids are released from adipose cells and mobilized for use. The process begins when levels of glucagon and adrenaline in the blood increase and these hormones bind to specific receptors on the surface of adipose cells. This binding action starts a cascade of reactions in the cell that results in the activation of yet another lipase that hydrolyzes triglyceride in the fat droplet to produce free fatty acids. These fatty acids are released into the circulatory system and delivered to skeletal and heart muscle as well as to the liver. In the blood the fatty acids are bound to a protein called serum albumin; in muscle tissue they are taken up by the cells and oxidized to carbon dioxide (CO_2) and water to produce energy, as described below. It is not clear whether a special transport mechanism is required for enabling free fatty acids to enter cells from the circulation.

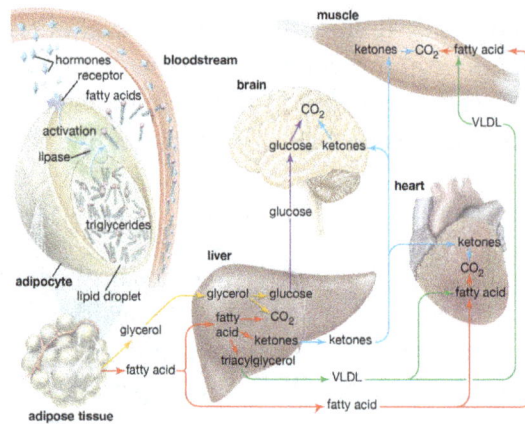

When hormones signal the need for energy, fatty acids and glycerol are released from triglycerides stored in fat cells (adipocytes) and are delivered to organs and tissues in the body.

The liver takes up a large fraction of the fatty acids. There they are in part resynthesized into triglycerides and are transported in VLDL lipoproteins to muscle and other tissues. A fraction is also converted to small ketone molecules that are exported via the circulation to peripheral tissues, where they are metabolized to yield energy.

Oxidation of Fatty Acids

Inside the muscle cell, free fatty acids are converted to a thioester of a molecule called coenzyme A, or CoA. (A thioester is a compound in which the linking oxygen in an ester is replaced by a sulfur atom.) Oxidation of the fatty acid–CoA thioesters actually takes place in discrete vesicular bodies called mitochondria. Most cells contain many mitochondria, each roughly the size of a bacterium, ranging from 0.5 to 10 m (micrometre; 1 m = one-millionth of a metre) in diameter; their size and shape differ depending on the cell type in which they occur. The mitochondrion is surrounded by a double membrane system enclosing a fluid interior space called the matrix. In the matrix are found the enzymes that convert the fatty acid–CoA thioesters into CO_2 and water (the chemical waste products of oxidation) and also adenosine triphosphate (ATP), the energy currency of living systems. The process consists of four sequential steps.

The first step is the transport of the fatty acid across the innermost of the two concentric mitochondrial membranes. The outer membrane is very porous so that the CoA thioesters freely permeate through it. The impermeable inner membrane is a different matter; here the fatty acid chains are transported across in the following way. On the cytoplasmic side of the membrane, an enzyme catalyzes the transfer of the fatty acid from CoA to a molecule of carnitine, a hydroxy amino acid. The carnitine ester is transported across the membrane by a transferase protein located in the membrane, and on the matrix side a second enzyme catalyzes the transfer of the fatty acid from carnitine back to CoA. The carnitine that is re-formed by loss of the attached fatty acid is transferred back to the cytoplasmic side of the mitochondrial membrane to be reused. The transfer of a fatty acid from the cytoplasm to the mitochondrial matrix thus occurs without the transfer of CoA itself from one compartment to the other. No energy is generated or consumed in this transport process, although energy is required for the initial formation of the fatty acid–CoA thioester in the cytoplasm.

The second step is the oxidation of the fatty acid to a set of two-carbon acetate fragments with thioester linkages to CoA. This series of reactions, known as β-oxidation, takes place in the matrix of the mitochondrion. Since most biological fatty acids have an even number of carbons, the number of acetyl-CoA fragments derived from a specific fatty acid is equal to one-half the number of carbons in the acyl chain. For example, palmitic acid (C_{16}) yields eight acetyl-CoA thioesters. In the case of rare unbranched fatty acids with an odd number of carbons, one three-carbon CoA ester is formed as well as the two-carbon acetyl-CoA thioesters. Thus, a C_{17} acid yields seven acetyl and one three-carbon CoA thioester. The energy in the successive oxidation steps is conserved by chemical reduction (the opposite of oxidation) of molecules that can subsequently be used to form ATP. ATP is the common fuel used in all the machinery of the cell (e.g., muscle, nerves, membrane transport systems, and biosynthetic systems for the formation of complex molecules such as DNA and proteins).

The two-carbon residues of acetyl-CoA are oxidized to CO_2 and water, with conservation of chemical energy in the form of $FADH_2$ and NADH and a small amount of ATP. This process is carried out in a series of nine enzymatically catalyzed reactions in the mitochondrial matrix space. The reactions form a closed cycle, often called the citric acid, tricarboxylic acid, or Krebs cycle.

The final stage is the conversion of the chemical energy in NADH and $FADH_2$ formed in the second and third steps into ATP by a process known as oxidative phosphorylation. All the participating enzymes are located inside the mitochondrial inner membrane—except one, which is trapped in the space between the inner and outer membranes. In order for the process to produce ATP, the inner membrane must be impermeable to hydrogen ions (H^+). In the course of oxidative phosphorylation, molecules of NADH and $FADH_2$ are subjected to a series of linked oxidation-reduction reactions. NADH and $FADH_2$ are rich in electrons and give up these electrons to the first member of the reaction chain. The electrons then pass down the series of oxidation-reduction reactions and in the last reaction reduce molecular oxygen (O_2) to water (H_2O). This part of oxidative phosphorylation is called electron transport.

The chemical energy available in these electron-transfer reactions is conserved by pumping H+ across the mitochondrial inner membrane from matrix to cytoplasm. Essentially an electrical battery is created, with the cytoplasm acting as the positive pole and the mitochondrial matrix as the negative pole. The net effect of electron transport is thus to convert the chemical energy of oxidation into the electrical energy of the transmembrane "battery." The energy stored in this battery is in turn used to generate ATP from adenosine diphosphate (ADP) and inorganic phosphate by the action of a complex enzyme called ATP synthase, also located on the inner mitochondrial membrane. It is interesting that a similar process forms the basis of photosynthesis—the mechanism by which green plants convert light energy from the Sun into carbohydrates and fats, the basic foods of both plants and animals. Many of the molecular details of the oxidative phosphorylation system are now known, but there is still much to learn about it and the equally complex process of photosynthesis.

The β-oxidation also occurs to a minor extent within small subcellular organelles called peroxisomes in animals and glyoxysomes in plants. In these cases fatty acids are oxidized to CO_2 and water, but the energy is released as heat. The biochemical details and physiological functions of these organelles are not well understood.

Regulation of Fatty Acid Oxidation

The rate of utilization of acetyl-CoA, the product of β-oxidation, and the availability of free fatty acids are the determining factors that control fatty acid oxidation. The concentrations of free fatty acids in the blood are hormone-regulated, with glucagon stimulating and insulin inhibiting fatty acid release from adipose tissue. The utilization in muscle of acetyl-CoA depends upon the activity of the citric acid cycle and oxidative phosphorylation—whose rates in turn reflect the demand for ATP.

In the liver the metabolism of free fatty acids reflects the metabolic state of the animal. In well-fed animals the liver converts excess carbohydrates to fatty acids, whereas in fasting animals fatty acid oxidation is the predominant activity, along with the formation of ketones. Although the details are not completely understood, it is clear that in the liver the metabolism of fatty acids is tightly linked to fatty acid synthesis so that a wasteful closed cycle of fatty acid synthesis from and metabolism back to acetyl-CoA is prevented.

Lipids in Biological Membranes

Biological membranes separate the cell from its environment and compartmentalize the cell interior. The various membranes playing these vital roles are composed of roughly equal weight percent protein and lipid, with carbohydrates constituting less than 10 percent in a few membranes. Although many hundreds of molecular species are present in any one membrane, the general organization of the generic components is known. All the lipids are amphipathic, with their hydrophilic (polar) and hydrophobic (nonpolar) portions located at separate parts of each molecule. As a result, the lipid components of membranes are arranged in what may be called a continuous bimolecular leaflet, or bilayer. The polar portions of the constituent molecules lie in the two bilayer faces, while the nonpolar portions constitute the interior of the bilayer. The lipid bilayer structure forms an impermeable barrier for essential water-soluble substances in the cell and provides the basis for the compartmentalizing function of biological membranes.

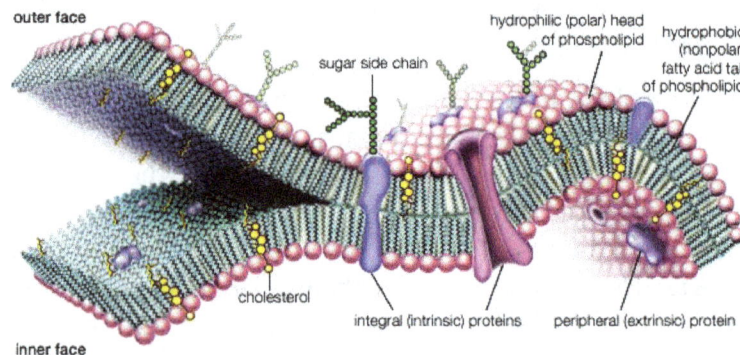

Molecular view of the cell membrane.

Intrinsic proteins penetrate and bind tightly to the lipid bilayer, which is made up largely of phospholipids and cholesterol and which typically is between 4 and 10 nanometers (nm; 1 nm = 10^{-9} metre) in thickness. Extrinsic proteins are loosely bound to the hydrophilic (polar) surfaces, which face the watery medium both inside and outside the cell. Some intrinsic proteins present sugar side chains on the cell's outer surface.

Some protein components are inserted into the bilayer, and most span this structure. These so-called integral, or intrinsic, membrane proteins have amino acids with nonpolar side chains at the interface between the protein and the nonpolar central region of the lipid bilayer. A second class of proteins is associated with the polar surfaces of the bilayer and with the intrinsic membrane proteins. The protein components are specific for each type of membrane and determine their predominant physiological functions. The lipid component, apart from its critical barrier function, is for the most part physiologically silent, although derivatives of certain membrane lipids can serve as intracellular messengers.

The most remarkable feature of the general biomembrane structure is that the lipid and the protein components are not covalently bonded to one another or to molecules of the other group. This sheetlike structure, formed only by molecular associations, is less than 10 nm in thickness but many orders of magnitude larger in its other two dimensions. Membranes are surprisingly strong mechanically, yet they exhibit fluidlike properties. Although the surfaces of membranes contain polar units, they act as an electric insulator and can withstand several hundred thousand volts without breakdown. Experimental and theoretical studies have established that the structure and these unusual properties are conferred on biological membranes by the lipid bilayer.

Composition of the Lipid Bilayer

Most biological membranes contain a variety of lipids, including the various glycerophospholipids such as phosphatidyl-choline, -ethanolamine, -serine, -inositol, and -glycerol as well as sphingo-myelin and, in some membranes, glycosphingolipids. Cholesterol, ergosterol, and sitosterol are sterols found in many membranes. The relative amounts of these lipids differ even in the same type of cell in different organisms, as shown in the table on the lipid composition of red blood cell membranes from different mammalian species. Even in a single cell, the lipid compositions of the membrane surrounding the cell (the plasma membrane) and the membranes of the various organelles within the cell (such as the microsomes, mitochondria, and nucleus) are different, as shown in the table on various membranes in a rat liver cell.

Organelle membrane lipid composition by weight percent of rat liver cells					
	membrane				
lipid	plasma membrane	microsome	inner mitochondria	outer mitochondria	nuclear
cholesterol	28.0	6.0	<1.0	6.0	5.1
phosphatidylcholine	31.0	55.20	37.9	42.70	58.30
sphingomyelin	16.6	3.7	00.8	4.1	3.0
phosphatidylethanolamine	14.3	24.00	38.3	28.60	21.50
phosphatidylserine	02.7	—	<1.0	<1.00	3.4
phosphatidylinositol	04.7	7.7	02.0	7.9	8.2
phosphatidic acid and cardiolipin	01.4	1.5	20.4	8.9	<1.00
lysophosphatidylcholine	01.3	1.9	00.6	1.7	1.4

Plasma membrane lipid composition by weight percent of mammalian red blood cells						
	species					
lipid	pig	human	cat	rabbit	horse	rat
cholesterol	26.8	26.0	26.8	28.9	24.5	24.7
phosphatidylcholine	13.9	17.5	18.7	22.3	22.0	31.8
sphingomyelin	15.8	16.0	16.0	12.5	07.0	08.6
phosphatidylethanolamine	17.7	16.6	13.6	21.0	12.6	14.4
phosphatidylserine	10.6	07.9	08.1	08.0	09.4	07.2
phosphatidylinositol	01.1	01.2	04.5	01.0	<0.2	02.3
phosphatidic acid	<0.2	00.6	00.5	01.0	<0.2	<0.2
lysophosphatidylcholine	00.5	00.9	<0.2	<0.2	00.9	02.6
glycosphingolipids	13.4	11.0	11.9	05.3	23.5	08.3

On the other hand, the lipid compositions of all the cells of a specific type in a specific organism at a given time in its life are identical and thus characteristic. During the life of an organism, there may be changes in the lipid composition of some membranes; the physiological significance of these age-related changes is unknown, however.

Physical Characteristics of Membranes

One of the most surprising characteristics of biological membranes is the fact that both the lipid and the protein molecules, like molecules in any viscous liquid, are constantly in motion. Indeed, the membrane can be considered a two-dimensional liquid in which the protein components ride like boats. However, the lipid molecules in the bilayer must always be oriented with their polar ends at the surface and their nonpolar parts in the central region of the bilayer. The bilayer structure thus has the molecular orientation of a crystal and the fluidity of a liquid. In this liquid-crystalline state, thermal energy causes both lipid and protein molecules to diffuse laterally and also to rotate about an axis perpendicular to the membrane plane. In addition, the lipids occasionally flip from one face of the membrane bilayer to the other and attach and detach from the surface of the bilayer at very slow but measurable rates. Although these latter motions are forbidden to intrinsic proteins, both lipids and proteins can exhibit limited bobbing motions. Within this seemingly random, dynamic mixture of components, however, there is considerable order in the plane of the membrane. This order takes the form of a "fluid mosaic" of molecular association complexes of both lipids and proteins in the membrane plane. The plane of the biological membrane is thus compartmentalized by domain structures much as the three-dimensional space of the cell is compartmentalized by the membranes themselves. The domain mosaics run in size from tens of nanometres (billionths of a metre) to micrometres (millionths of a metre) and are stable over time intervals of nanoseconds to minutes. In addition to this in-plane domain structure, the two lipid monolayers making up the membrane bilayer frequently have different compositions. This asymmetry, combined with the fact that intrinsic membrane proteins do not rotate about an axis in the membrane plane, makes the two halves of the bilayer into separate domains.

An interesting class of proteins is attached to biological membranes by a lipid that is chemically linked to the protein. Many of these proteins are involved in intra- and intercellular signaling. In some cases defects in their structure render the cells cancerous, presumably because growth-limiting signals are blocked by the structural error.

Intracellular and Extracellular Messengers

In multicellular organisms (eukaryotes), the internal mechanisms that control and coordinate basic biochemical reactions are connected to other cells by means of nerves and chemical "messengers." The overall process of receiving these messages and converting the information they contain into metabolic and physiological effects is known as signal transduction. Many of the chemical messengers are lipids and are thus of special interest here. There are several types of external messengers. The first of these are hormones such as insulin and glucagon and the lipids known collectively as steroid hormones. A second class of lipid molecules is eicosanoids, which are produced in tissues and elicit cellular responses close to their site of origin. They are produced in very low levels and are turned over very rapidly (in seconds). Hormones have sites of action that are remote from their cells of origin and remain in the circulation for long periods (minutes to hours).

Steroid Hormones

Lipid hormones invoke changes in gene expression; that is, their action is to turn on or off the instructions issued by deoxyribonucleic acid (DNA) to produce proteins that regulate the biosynthesis of other important proteins. Steroids are carried in the circulation bound singly to specific carrier proteins that target them to the cells in particular organs. After permeating the external membranes of these cells, the steroid interacts with a specific carrier protein in the cytoplasm. This soluble complex migrates into the cell nucleus, where it interacts with the DNA to activate or repress transcription, the first step in protein biosynthesis.

All five major classes of steroid hormones produced from cholesterol contain the characteristic five rings of carbon atoms of the parent molecule. Progestins are a group of steroids that regulate events during pregnancy and are the precursors of the other steroid hormones. The glucocorticoids, cortisol, and corticosterones promote the biosynthesis of glucose and act to suppress inflammation. The mineralocorticoids regulate ion balances between the interior and the exterior of the cell. Androgens regulate male sexual characteristics, and estrogens perform an analogous function in females. The target organs for these hormones are listed in the table.

Organs affected by steroid hormones	
hormone class	target organs
glucocorticoids	liver, retina, kidney, oviduct, pituitary
estrogens	oviduct, liver
progesterone	oviduct, uterus
androgens	prostate, kidney, oviduct

Eicosanoids

Three types of locally acting signaling molecules are derived biosynthetically from C20 polyunsaturated fatty acids, principally arachidonic acid. Twenty-carbon fatty acids are all known collectively as eicosanoic acids. The three chemically similar classes are prostaglandins, thromboxanes, and leukotrienes. The eicosanoids interact with specific cell surface receptors to produce a variety of different effects on different tissues, but generally they cause inflammatory responses and changes in blood pressure, and they also affect the clotting of blood. Little is known about how these effects are produced within the cells of target tissues. However, it is known that aspirin and other anti-inflammatory drugs inhibit either an enzyme in the biosynthesis pathway or the eicosanoid receptor on the cell surface.

Intracellular Second Messengers

With the exception of the steroid hormones, most hormones such as insulin and glucagon interact with a receptor on the cell surface. The activated receptor then generates so-called second messengers within the cell that transmit the information to the biochemical systems whose activities must be altered to produce a particular physiological effect. The magnitude of the end effect is generally proportional to the concentration of the second messengers.

An important intracellular second-messenger signaling system, the phosphatidylinositol system, employs two second-messenger lipids, both of which are derived from phosphatidylinositol. One is diacylglycerol (diglyceride), the other is triphosphoinositol. In this system a membrane receptor acts upon an enzyme, phospholipase C, located on the inner surface of the cell membrane. Activation of this enzyme by one of the agents listed in the table causes the hydrolysis of a minor membrane phospholipid, phosphatidylinositol bisphosphate. Without leaving the membrane bilayer, the diacylglycerol next activates a membrane-bound enzyme, protein kinase C, that in turn catalyzes the addition of phosphate groups to a soluble protein. This soluble protein is the first member of a reaction sequence leading to the appropriate physiological response in the cell. The other hydrolysis product of phospholipase C, triphosphoinositol, causes the release of calcium from intracellular stores. Calcium is required, in addition to triacylglycerol, for the activation of protein kinase C.

Tissue affected by phosphoinositide second-messenger system		
Extracellular signal	target tissue	cellular response
Acetylcholine	pancreas pancreas (islet cells) smooth muscle	amylase secretion insulin release contraction
Vasopressin	liver kidney	glycogenolysis
Thrombin	blood platelets	platelet aggregation
Antigens	lymphoblasts mast cells	DNA synthesis histamine secretion
Growth factors	fibroblasts	DNA synthesis
Spermatozoa	eggs (sea urchin)	fertilization
Light	photoreceptors (horseshoe crab)	phototransduction
Thyrotropin-releasing hormone	pituitary anterior lobe	prolactin secretion

Fatty Acids

Fatty acids are the part of complex lipids which play a number of key roles in metabolism – major metabolic fuel (storage and transport of energy), as essential components of all membranes, and as gene regulators.

dietary lipids provide polyunsaturated fatty acids (PUFAs) that are precursors of powerful locally acting metabolites, i.e. the eicosanoids. As part of complex lipids, fatty acids are also important for thermal and electrical insulation, and for mechanical protection. Moreover, free fatty acids and their salts may function as detergents and soaps owing to their amphipathic properties and the formation of micelles.

Structure of Fatty Acid

Fatty acids are carbon chains with a methyl group at one end of the molecule (designated omega, o) and a carboxyl group at the other end (The carbon atom next to the carboxyl group is called the a carbon, and the subsequent one the b carbon. The letter n is also often used instead of the Greek o to indicate the position of the double bond closest to the methyl end. The systematic nomenclature for fatty acids may also indicate the location of double bonds with reference to the carboxyl group (D).

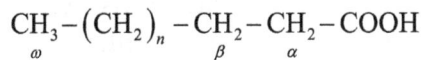

$$\underset{\omega}{CH_3} - (CH_2)_n - \underset{\beta}{CH_2} - \underset{\alpha}{CH_2} - COOH$$

Fatty acids may be named according to systematic or trivial nomenclature. One systematic way to describe fatty acids is related to the methyl (ω) end. This is used to describe the position of double bonds from the end of the fatty acid. The letter n is also often used to describe the o position of double bonds.

Saturated Fatty Acids

Saturated fatty acids are 'filled' (saturated) with hydrogen. Most saturated fatty acids are straight hydrocarbon chains with an even number of carbon atoms. The most common fatty acids contain 12–22 carbon atoms.

Unsaturated Fatty Acids

Monounsaturated fatty acids have one carbon–carbon double bond, which can occur in different positions. The most common monoenes have a chain length of 16–22 and a double bond with the cis configuration. This means that the hydrogen atoms on either side of the double bond are oriented in the same direction. Trans isomers may be produced during industrial processing (hydrogenation) of unsaturated oils and in the gastrointestinal tract of ruminants. The presence of a double bond causes restriction in the mobility of the acyl chain at that point. The cis configuration gives a kink in the molecular shape and cis fatty acids are thermodynamically less stable than the trans forms. The cis fatty acids have lower melting points than the trans fatty acids or their saturated counterparts.

In polyunsaturated fatty acids (PUFAs) the first double bond may be found between the third and the fourth carbon atom from the ω carbon; these are called ω-3 fatty acids. If the first double bond is between the sixth and seventh carbon atom, then they are called ω-6 fatty acids. The double bonds in PUFAs are separated from each other by a methylene grouping.

ω-characteristics	Methyl end	Carboxyl end	Saturation	Δ-characteristics
Stearic 18:0		COOH	Saturate	18:0
Oleic 18:1, ω-9		COOH	Monoene	18:1 Δ9
Linoleic 18:2, ω-6		COOH	Polyene	18:2 Δ9,12
α-Linolenic 18:3, ω-3		COOH	Polyene	18:3 Δ9,12,15
EPA 20:5, ω-3		COOH	Polyene	20:5 Δ5,8,11,14,17
DHA 22:6, ω-3		COOH	Polyene	20:6 Δ4,7,10,13,16,19

Figure: Structure of different unbranched fatty acids with a methyl end and a carboxyl (acidic) end.

Stearic acid is a trivial name for a saturated fatty acid with 18 carbon atoms and no double bonds (18:0). Oleic acid has 18 carbon atoms and one double bond in the ω-9 position (18:1 ω-9), whereas eicosapentaenoic acid (EPA), with multiple double bonds, is represented as 20:5 ω-3. This numerical scheme is the systematic nomenclature most commonly used. It is also possible to describe fatty acids systematically in relation to the acidic end of the fatty acids; symbolized D (Greek delta) and numbered 1. All unsaturated fatty acids are shown with cis configuration of the double bonds. DHA, docosahexaenoic acid.

PUFAs, which are produced only by plants and phytoplankton, are essential to all higher organisms, including mammals and fish. ω-3 and ω-6 fatty acids cannot be interconverted, and both are essential nutrients. PUFAs are further metabolized in the body by the addition of carbon atoms and by desaturation (extraction of hydrogen). Mammals have desaturases that are capable of removing hydrogens only from carbon atoms between an existing double bond and the carboxyl group. b-oxidation of fatty acids may take place in either mitochondria or peroxisomes.

Major Fatty Acids

Fatty acids represent 30–35% of total energy intake in many industrial countries and the most important dietary sources of fatty acids are vegetable oils, dairy products, meat products, grain and fatty fish or fish oils.

The most common saturated fatty acid in animals, plants and microorganisms is palmitic acid (16:0). Stearic acid (18:0) is a major fatty acid in animals and some fungi, and a minor component in most plants.Myristic acid (14:0) has a widespread occurrence, occasionally as a major component. Shorter-chain saturated acids with 8–10 carbon atoms are found in milk and coconut triacylglycerols.

Oleic acid (18:1 ω-9) is the most common monoenoic fatty acid in plants and animals. It is also found in microorganisms. Palmitoleic acid (16:1 ω-7) also occurs widely in animals, plants and microorganisms, and is a major component in some seed oils.

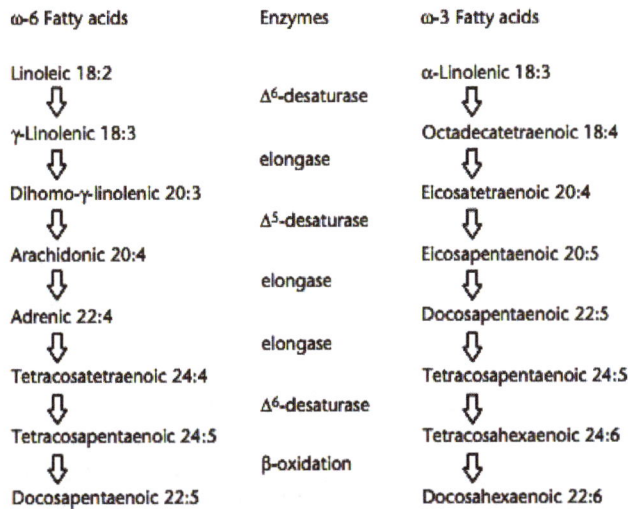

ω-6 Fatty acids	Enzymes	ω-3 Fatty acids
Linoleic 18:2		α-Linolenic 18:3
⇩	Δ^6-desaturase	⇩
γ-Linolenic 18:3		Octadecatetraenoic 18:4
⇩	elongase	⇩
Dihomo-γ-linolenic 20:3		Eicosatetraenoic 20:4
⇩	Δ^5-desaturase	⇩
Arachidonic 20:4		Eicosapentaenoic 20:5
⇩	elongase	⇩
Adrenic 22:4		Docosapentaenoic 22:5
⇩	elongase	⇩
Tetracosatetraenoic 24:4		Tetracosapentaenoic 24:5
⇩	Δ^6-desaturase	⇩
Tetracosapentaenoic 24:5		Tetracosahexaenoic 24:6
⇩	β-oxidation	⇩
Docosapentaenoic 22:5		Docosahexaenoic 22:6

Figure: Synthesis of ω-3 and ω-6 polyunsaturated fatty acids (PUFAs).

There are two families of essential fatty acids that are metabolized in the body as shown in this figure. Retroconversion, e.g. DHA → EPA also takes place.

Linoleic acid (18:2 ω-6) is a major fatty acid in plant lipids. In animals it is derived mainly from dietary plant oils. Arachidonic acid (20:4 -6) is a major component of membrane phospholipids throughout the animal kingdom, but very little is found in the diet. a-Linolenic acid (18:3 ω-3) is found in higher plants (soyabean oil and rape seed oils) and algae. Eicosapentaenoic acid (EPA; 20:5 ω-3) and docosahexaenoic acid (DHA; 22:6 ω-3) are major fatty acids of marine algae, fatty fish and fish oils; for example, DHA is found in high concentrations, especially in phospholipids in the brain, retina and testes.

Metabolism of Fatty Acids

An adult consumes approximately 85 g of fat daily, most of it as triacylglycerols. During digestion, free fatty acids

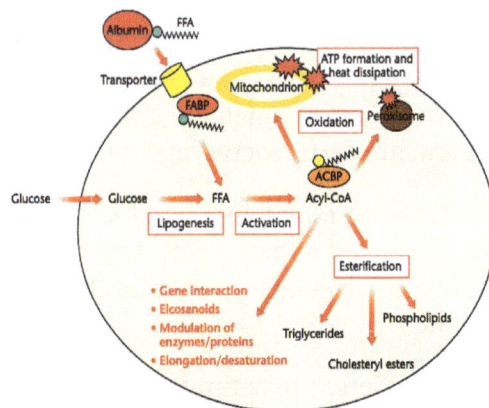

Figure: Metabolism of fatty acids.

Free fatty acids (FFA) are taken up into cells mainly by protein carriers in the plasma membrane and transported intracellularly via fatty acid-binding proteins (FABP). FFA are activated (acyl-

CoA) before they can be shuttled via acyl-CoA binding protein (ACBP) to mitochondria or per-oxisomes for b-oxidation (formation of energy as ATP and heat), or to endoplasmic reticulum for esterification to different lipid classes. Acyl-CoA or certain FFA may bind to transcription factors that regulate gene expression or may be converted to signalling molecules (eicosanoids). Glucose may be transformed to fatty acids if there is a surplus of glucose/energy in the cells.

(FFA) and monoacylglycerols are released and absorbed in the small intestine. In the intestinal mucosa cells, FFA are re-esterified to triacylglycerols, which are transported via lymphatic vessels to the circulation as part of chylomicrons. In the circulation, fatty acids are transported bound to albumin or as part of lipoproteins.

FFA are taken up into cells mainly by protein transporters in the plasma membrane and are trans-ported intracellularly via fatty acid-binding proteins (FABP). FFA are then activated (acyl-CoA) before they are shuttled via acyl-CoA-binding protein (ACBP) to mitochondria or peroxisomes for b-oxidation (and formation of energy as ATP and heat) or to endoplasmic reticulum for esterifi-cation to different classes of lipid. Acyl-CoA or certain FFA may bind to transcription factors that regulate gene expression or may be converted to signal molecules (eicosanoids). Glucose may be transformed to fatty acids (lipogenesis) if there is a surplus of glucose/energy in the cells.

Properties of Fatty Acids

Physical Properties

Fatty acids are poorly soluble in water in their undissociated (acidic) form, whereas they are rel-atively hydrophilic as potassium or sodium salts. Thus, the actual water solubility, particularly of longer-chain acids, is often very difficult to determine since it is markedly influenced by pH, and also because fatty acids have a tendency to associate, leading to the formation of monolayers or mi-celles. The formation of micelles in aqueous solutions of lipids is associated with very rapid chang-es in physical properties over a limited range of concentration. The point of change is known as the critical micellar concentration (CMC), and exemplifies the tendency of lipids to associate rather than remain as single molecules. The CMC is not a fixed value but represents a small concentration range that is markedly affected by the presence of other ions and by temperature.

Fatty acids are easily extracted with nonpolar solvents from solutions or suspensions by lowering the pH to form the uncharged carboxyl group. In contrast, raising the pH increases water solubil-ity through the formation of alkali metal salts, which are familiar as soaps. Soaps have important properties as association colloids and are surfaceactive agents.

The influence of a fatty acid's structure on its melting point is such that branched chains and cis double bonds will lower the melting point compared with that of equivalent saturated chains. In addition, the melting point of a fatty acid depends on whether the chain is even- or oddnumbered; the latter have higher melting points.

Saturated fatty acids are very stable, whereas unsaturated acids are susceptible to oxidation: the more double bonds, the greater the susceptibility. Thus, unsaturated fatty acids should be han-dled under an atmosphere of inert gas and kept away from oxidants and compounds giving rise to formation of free radicals. Antioxidants may be very important in the prevention of potentially harmful attacks on acyl chains in vivo.

Mechanisms of Action

The different mechanisms by which fatty acids can influence biological systems are outlined in figure.

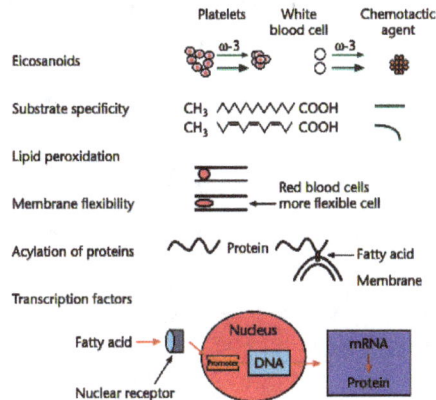

Figure: Mechanisms of action for fatty acids. Thromboxanes formed in blood platelets promote aggregation (clumping) of blood platelets. Leukotrienes in white blood cells act as chemotactic agents (attracting other white blood cells).

Eicosanoids

Eikosa means 'twenty' in Greek, and denotes the number of carbon atoms in the PUFAs that act as precursors of eicosanoids. These signalling molecules are called leukotrienes, prostaglandins, thromboxanes, prostacyclins, lipoxins and hydroperoxy fatty acids. Eicosanoids are important for several cellular functions such as platelet aggregability (ability to clump and fuse), chemotaxis (movement of blood cells) and cell growth. Eicosanoids are rapidly produced and degraded in cells where they execute their effects. Different cell types produce various types of eicosanoids with different biological effects. For example, platelets mostly make thromboxanes, whereas endothelial cells mainly produce prostacyclins. Eicosanoids from the o-3 PUFAs are usually less potent than eicosanoids derived from the o-6 fatty acids.

Substrate Specificity

Fatty acids have different abilities to interact with enzymes or receptors, depending on their structure. For example, EPA is a poorer substrate than all other fatty acids for esterification to cholesterol and diacylglycerol. Some ω-3 fatty acids are preferred substrates for certain desaturases. The preferential incorporation of ω-3 fatty acids into some phospholipids occurs because ω-3 fatty acids are preferred substrates for the enzymes responsible for phospholipid synthesis. These examples of altered substrate specificity of ω-3 PUFA for certain enzymes illustrate why EPA and DHA are mostly found in certain phospholipids.

Membrane Fluidity

When large amounts of vhery long-chain o-3 fatty acids are ingested, there is a high incorporation of EPA and DHA into membrane phospholipids. An increased amount of ω-3 PUFA may change the physical characteristics of the membranes. Altered fluidity may lead to changes of membrane protein functions. The very large amount of DHA in phosphatidylethanolamine and phosphatidylserine in certain areas of the retinal rod outer segments is probably crucial for the function of membrane phospholipids in light transduction, because these lipids are located close to the rhodopsin molecules. It has been shown that the flexibility of membranes from blood cells is increased in animals fed fish oil, and this might be important for the microcirculation. Increased incorporation of very longchain ω-3 PUFAs into plasma lipoproteins changes the physical properties of low-density lipoproteins (LDL), lowering the melting point of core cholesteryl esters.

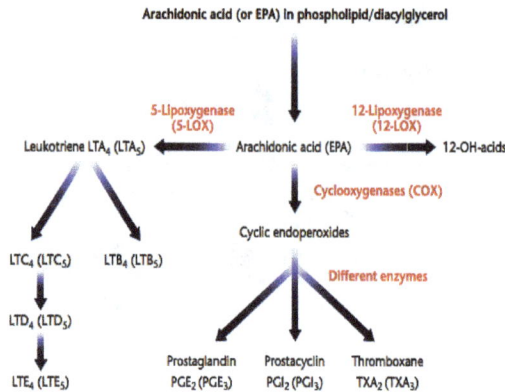

Figure: Synthesis of eicosanoids from arachidonic acid or eicosapentaenoic acid (EPA).

Fatty acid	AA	EPA	AA	EPA	AA	EPA
Enzyme		Cyclooxygenase			Lipoxygenase	
Cell type	Platelets		Endothelial cells		Leucocytes	
Eicosanoids	TXA_2	TXA_3	PGI_2	PGI_3	LTB_4	LTB_5

Biological effect

	AA	EPA	AA	EPA	AA	EPA
Aggregation	+++	+				
Antiaggregation			+++	+++		
Vasoconstriction	+++					
Vasodilatation			+++	+++		
Chemotaxis					+++	+

Figure: Biological effects of eicosanoids derived from arachidonic acid (AA; 20:4 ω-6) or eicosapentaenoic acid (EPA; 20:5 ω-3). TX, thromboxane; PG, prostaglandin, LT, leukotriene.

Lipid Peroxidation

Lipid peroxidation products may act as biological signals. One of the major concerns with intake of PUFAs has been their high degree of unsaturation, and therefore the possibility that they might facilitate peroxidation of LDL. Peroxidized LDL might be endocytosed by macrophages and initiate development of atherosclerosis. Oxidatively modified LDL has been found in atherosclerotic lesions, and LDL rich in oleic acid was found to be more resistant to oxidative modification than LDL enriched with o-6 fatty acids in rabbits. Although some of the published data are conflicting, several well-performed studies indicate small or no harmful effects of o-3 fatty acids. It should be recalled from the results of epidemiological studies that the dietary intake of saturated fatty acids, transfatty acids and cholesterol is strongly correlated with development of coronary heart disease, whereas intake of PUFAs is related to reduced incidence of coronary heart disease. Several studies suggest that it is important that the proper amount of antioxidants is included in the diet with the PUFA to decrease the risk of lipid peroxidation

Acylation of Proteins

Some proteins are acylated with stearic (18:0), palmitic (16:0) or myristic (14:0) acids. This acylation of proteins is important for anchoring certain proteins in membranes or for folding of the proteins, and is crucial for the function of these proteins. Although the saturated fatty acids are most commonly covalently linked to proteins, PUFA may also acylate proteins.

Gene Interactions

Fatty acids or their derivatives (acyl-CoA or eicosanoids) may interact with nuclear receptor proteins that bind to certain regulatory regions of DNA and thereby alter transcription of these genes. The combined fatty acid– receptor complex may function as a transcription factor. The first example of this was the peroxisome proliferatoractivated receptor (PPAR). Natural fatty acids are weak activators of PPAR, and this may be explained by the rapid oxidation of fatty acids. If fatty acids are blocked from being oxidized, they may be more potent stimulators of PPAR than natural fatty acids. Fatty acids may also influence expression of several glycolytic and lipogenic genes independently of PPAR. It has been demonstrated that one eicosanoid derived from arachidonic acid, prostaglandin J_2 (PGJ$_2$), binds to PPARγ, which is an important transcription factor found in adipose tissue. PUFA may also influence proliferation of white blood cells, together with the cells' tendency to die by programmed cell death (apoptosis) or necrosis. Thus, fatty acids may be important for regulation of gene transcription and thereby regulate metabolism, cell proliferation and cell death.

Biological Effects

Replacement of saturated fat with monounsaturated and polyunsaturated fat (especially ω-6 PUFA) decreases the plasma concentration of total and LDL cholesterol. The mechanism for these effects may be increased uptake of LDL particles from the circulation by the liver.

Table: Effect of fatty acids on plasma and LDL cholesterola		
	Δ Cholesterol (mmol L^{-1})	Δ LDL cholesterol (mmol L^{-1})
12:0	+0.01	+0.01
14:0	+0.12	+0.071
16:0	+0.057	+0.047
Trans Marine[b]	+0.039	+0.043
Trans Veg	+0.031	+0.025
18:1	-0.0044	-0.0044
18:2/3	-0.017	-0.017

Trans Marine, trans fatty acids of marine origin; trans Veg, transfatty acids of vegetable origin.

Dietary marine ω-3 fatty acids (EPA and DHA) decrease plasma triacylglycerol levels by reducing production and enhancing clearance of triacylglycerol-rich lipoproteins. In addition to effects on plasma lipids, dietary fatty acids can influence metabolic, immunological and cardiovascular events in numerous ways. For instance, saturated fat may negatively affect several factors related to cardiovascular diseases and atherosclerosis, whereas very longchain ω-3 PUFAs may exert several beneficial effects on the cardiovascular system. Briefly, ω-3 PUFAs decrease platelet and leucocyte reactivity, inhibit lymphocyte proliferation, and slightly decrease blood pressure. ω-3 PUFAs may also beneficially influence vessel wall characteristics and blood rheology, prevent ventricular arrhythmias and improve insulin sensitivity. ω-6 PUFAs (mainly linoleic acid, 18:2 ω-6) also have many beneficial effects with respect to cardiovascular diseases.

The essential ω-3 and ω-6 fatty acids are important for fetal growth and development, in particular for the central nervous system, affecting visual acuity as well as cognitive function. Lack of essential fatty acids also promotes skin inflammations and delays wound healing.

EPA and DHA have consistently been shown to inhibit proliferation of certain cancer cell lines in vitro and to reduce progression of these tumours in animal experiments. However, it is still unclear whether human cancer development is beneficially influenced by fatty acids.

Fatty Acid Oxidation and Synthesis

Fatty acid β-oxidation is a multi step process by which fatty acids are broken down by various tissues to produce energy. It involves first getting the fatty acid into the cytosol and then transferring it to the mitochondria where β-oxidation takes place. On being released from chylomicrons or very-low-density lipoproteins (VLDLs) by lipoprotein lipase in the capillary beds in the body, free fatty acids primarily enter the cell via fatty acid transporters on the cell surface. Fatty acid transporters include fatty acid translocase (FAT/CD36), tissue specific fatty acid transport proteins (FATP), and plasma membrane bound fatty acid binding protein (FABPpm). The processes involved in fatty acid uptake and oxidation are summarized in figure.

Once inside the cell fatty acids are activated by conjugation with coenzyme A (CoA) in a reaction catalyzed by acyl-CoA synthetase (thiokinase). This enzyme is associated with the endoplasmic reticulum and outer mitochondrial membrane and requires ATP. The ATP is cleaved to AMP plus PPi in this reaction. A second enzyme inorganic pyrophosphatase then cleaves the PPi to 2 molecules of Pi which helps to drive the acylation reaction to completion. Fatty acid oxidation and fatty acid synthesis require the acyl group to be covalently attached to either coenzyme A (oxidation) or acyl carrier protein (synthesis). In both cases their carboxyl groups are covalently linked to the terminal cysteine of a phosphopantetheine group, with CoA being the source of the phosphopantetheine group attached to ACP.

While fatty acid activation occurs in the cytosol, β-oxidation occurs inside the mitochondrion. Hence the long chain acyl-CoA is transported across the impermeable mitochondrial membrane after conversion of the long chain acyl-CoA to long chain acylcarnitine by the enzyme carnitine palmitoyltransferase-1 (CPT1). CPT1, resides on the inner surface of the outer mitochondrial membrane, and is a major site of regulation of mitochondrial fatty acid uptake. The fatty acylcarnitine moiety is transported across the inner mitochondrial membrane via the transport protein carnitine translocase (CAT), which exchanges long chain acylcarnitines for carnitine. An inner mitochondrial membrane carnitine palmitoyltransferase-2 (CPT2) then converts the long chain acylcarnitine back to long chain acyl-CoA which is now ready to be metabolized by the β-oxidation pathway.

β-oxidation

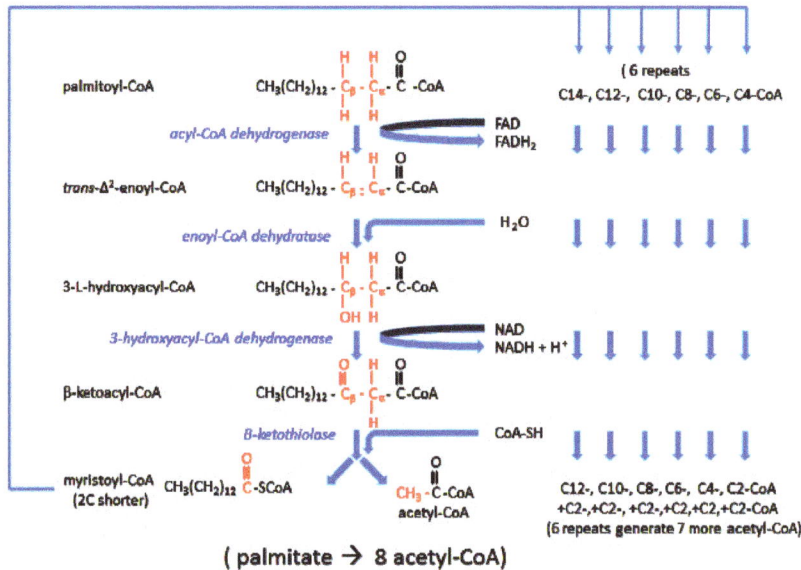

Figure: The fatty acid oxidation cycle. The example shown is for the oxidation of the C16 fatty acid palmitate.

Each cycle of four reactions generates one acetyl-CoA and an acyl-CoA that is two C units shorter. As shown in figure, there are four enzymes involved in β-oxidation: acyl-CoA dehydrogenase, enoyl-CoA hydratase, hydroxy acyl-CoA dehydrogenase, and ketoacyl-CoA thiolase. In the first step acyl-CoA dehydrogenase creates a double bond between the second and third carbons down from the CoA group on acyl-CoA and in the process produces one molecule of FADH2. In the second step, enoyl-CoA hydratase adds a water molecule by removing the double bond just formed, adding a hydroxyl group to the third carbon down from the CoA group and a hydrogen to the second carbon down from the CoA group.

In the third step hydroxyacyl-CoA dehydrogenase removes the hydrogen in the hydroxyl group just attached and in the process produces a molecule of NADH. In the final step, ketoacyl-CoA thiolase splits off the terminal acetyl-CoA moiety by attaching a new CoA group to the third carbon upstream of the original CoA group, resulting in the formation of two molecules, acetyl-CoA and an acyl-CoA that is now two carbons shorter. β-oxidation of long chain acyl CoA results in the production of one acetyl-CoA from each cycle of fatty acid β-oxidation. This acetyl-CoA then enters the mitochondrial tricarboxylic acid cycle (TCA). The NADH and FADH2 produced by both fatty acid β-oxidation and the TCA cycle are used by the electron transport chain to produce ATP.

A list of saturated fatty acids, their formulae and alternative terminology is shown in table.

Table: List of some saturated fatty acids.

Common Name	Systematic Name	Structural Formula	Lipid Numbers
Propionic acid	Propanoic acid	CH3CH2COOH	C3:0
Butyric acid	Butanoic acid	CH3(CH2)2COOH	C4:0
Valeric acid	Pentanoic acid	CH3(CH2)3COOH	C5:0

Common Name	Systematic Name	Structural Formula	Lipid Numbers
Caproic acid	Hexanoic acid	$CH_3(CH_2)_4COOH$	C6:0
Enanthic acid	Heptanoic acid	$CH_3(CH_2)_5COOH$	C7:0
Caprylic acid	Octanoic acid	$CH_3(CH_2)_6COOH$	C8:0
Pelargonic acid	Nonanoic acid	$CH_3(CH_2)_7COOH$	C9:0
Capric acid	Decanoic acid	$CH_3(CH_2)_8COOH$	C10:0
Undecylic acid	Undecanoic acid	$CH_3(CH_2)_9COOH$	C11:0
Lauric acid	Dodecanoic acid	$CH_3(CH_2)_{10}COOH$	C12:0
Tridecylic acid	Tridecanoic acid	$CH_3(CH_2)_{11}COOH$	C13:0
Myristic acid	Tetradecanoic acid	$CH_3(CH_2)_{12}COOH$	C14:0
Pentadecylic acid	Pentadecanoic acid	$CH_3(CH_2)_{13}COOH$	C15:0
Palmitic acid	Hexadecanoic acid	$CH_3(CH_2)_{14}COOH$	C16:0
Margaric acid	Heptadecanoic acid	$CH_3(CH_2)_{15}COOH$	C17:0
Stearic acid	Octadecanoic acid	$CH_3(CH_2)_{16}COOH$	C18:0

Short- and medium-chain fatty acids of between 4-12 carbon atoms in length are oxidized exclusively in the mitochondria whereas long-chain fatty acids C12-C16 are oxidized in both the mitochondria and peroxisomes. Longer chain fatty acids (C17–C26) are preferentially oxidized in the peroxisomes rather than in mitochondria with cerotic acid (a 26:0 fatty acid) being solely oxidized in this organelle. The peroxisomes also metabolize di– and trihydroxycholestanoic acids (bile acid intermediates); long-chain dicarboxylic acids that are produced by ω-oxidation of long-chain monocarboxylic acids; pristanic acid and certain polyunsaturated fatty acids (PUFAs).

Energy Yield from Fatty Acid Oxidation

Fat is the body's preferred way of storing energy owing to its high energy density (9 kCal/g compared to 4 kCal/g for carbohydrates and proteins). The amount of energy (ATP) generated per mole of stearic acid (C18) is 120 ATP molecules. In contrast, the complete oxidation of three molecules of glucose (3 x C6 = C18) via glycolysis to pyruvate, decarboxylation to acetyl-CoA and oxidation of acetylCoA via the TCA cycle and electron transfer chain generates only 90 ATP, 33% less than that generated from stearate.

Enzymes Involved in Fatty Acid Synthesis

Fatty acid synthesis is not simply a reversal of the oxidative pathway. The acyl intermediates are similar but the pathway consists of a new set of reactions, exemplifying the principle that synthetic and degradative pathways are usually distinct. In both pathways the acyl intermediates are attached to a prosthetic group which is CoA for oxidation and acyl carrier protein (ACP) for fatty acid synthesis. Both CoA and ACP have phosphopantetheine as their reactive units to which the fatty acid moiety is attached.

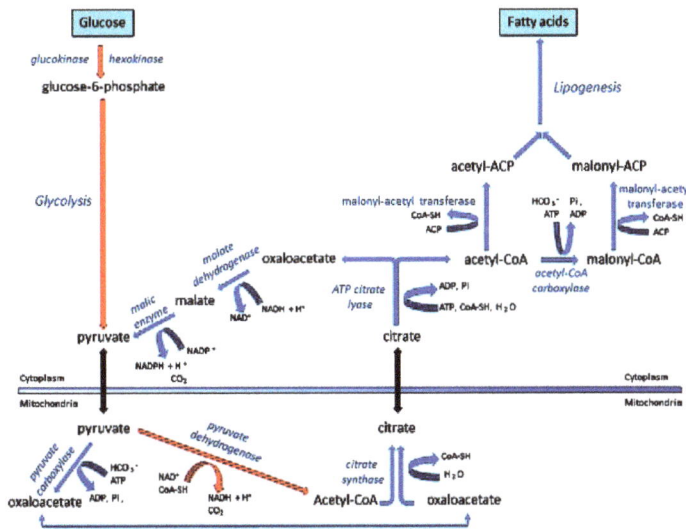

Figure: Linking glycolysis to fatty acid synthesis involves mitochondrial and cytosolic enzymes.

Pyruvate, the end product of glycolysis is transported to the mitochondria and decarboxylated to form acetyl-CoA, the key building block of fatty acid synthesis. Acetyl-CoA combines with oxaloacetate for transport to the cytoplasm as citrate. There acetyl- CoA is regenerated by ATP citrate lyase and activated to malonyl-CoA. Before the fatty acid synthesis cycle can commence the acetyl- and malonyl- groups are transferred from their linkage with CoA to a larger prosthetic group, acyl carrier protein (ACP). The oxaloacetate lost from the mitochondrial pool through the formation and transport of citrate is returned as pyruvate after reduction to malate and decarboxylation. An overview of fatty acid synthesis is shown in figure. In liver cells, glycolysis converts excess glucose to pyruvate. Each molecule of glucose (a six carbon sugar) is catabolised to form two molecules of pyruvate (a 3-carbon sugar) which is transported into the mitochondria via the transport protein pyruvate translocase. There it is decarboxylated by the enzyme pyruvate dehydrogenase to form acetyl-CoA, which combines with oxaloacetate to form the tricarboxylic acid, citrate which may be oxidized further in the mitochondrion via the tricarboxylic acid cycle (Krebs Cycle). Excess citrate that is not oxidized via the tricarboxylic acid cycle is exported to the cytosol for fatty acid synthesis. There citrate is cleaved into oxaloacetate and acetyl-CoA by the enzyme ATP citrate lyase. This enzyme is an important link between the metabolism of carbohydrates and the production of fatty acids as the acetyl-CoA so produced is the building block for the synthesis of fatty acids and cholesterol.

The oxaloacetate employed in the transfer of acetyl groups to the cytosol needs to be returned to the mitochondria. Since the inner mitochondrial membrane is impermeable to oxaloacetate a series of bypass reactions occurs in the cytosol: malate dehydrogenase reduces oxaloacetate to malate and the NADP-linked enzyme, malate enzyme, decarboxylates the newly formed malate to pyruvate, generating one molecule of NADH+ H+ in the process. The pyruvate formed in this reaction readily enters the mitochondria, where it is carboxylated to oxaloacetate by pyruvate carboxylase. The NADH+ H+ generated is required for the reductive steps in fatty acid synthesis.

Fatty acid synthesis starts with the carboxylation of acetyl-CoA to malonyl-CoA by the enzyme acetyl-CoA carboxylase (ACC). This irreversible reaction, which requires one molecule of ATP is the committed step in fatty acid synthesis and ACC is the essential regulatory enzyme for fatty acid metabolism. The intermediates involved in the subsequent steps in fatty acid synthesis are linked to an acyl carrier protein (ACP) via the sulfhydryl terminus of a phosphopantetheine group, which, in turn, is attached to a serine residue of the acyl carrier protein. ACP is a single polypeptide chain of 77 residues and can be regarded as a giant prosthetic group, a "macro CoA.". In mammals the generation of acetyl-ACP and malonyl-ACP are catalysed by a single bifunctional protein domain, malonyl-acetyl transferase (MAT).

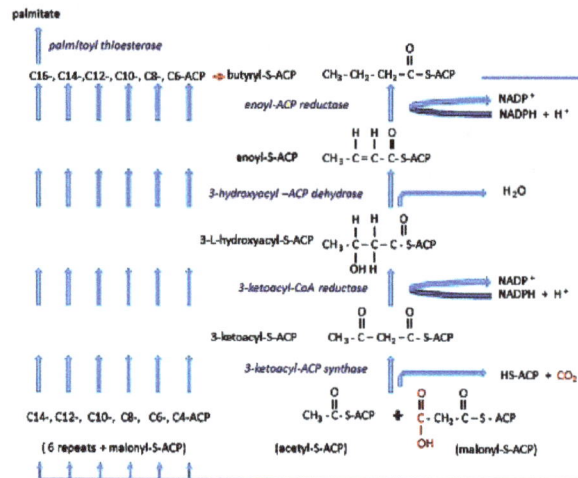

Figure: The fatty acid synthesis cycle.

The example shown is for the generation of the C16 fatty acid palmitate. Each cycle of four reactions consumes one malonyl-ACP and produces an acyl-ACP that is two C units longer. Seven cycles results in the production of the C16 intermediate palmitoyl-ACP which is hydrolysed to palmitate and ACP by the enzyme palmitoyl thioesterase. Further elongation and the insertion of double bonds are carried out by other enzyme systems. acid synthesis and ACC is the essential regulatory enzyme for fatty acid metabolism. The intermediates involved in the subsequent steps in fatty acid synthesis are linked to an acyl carrier protein (ACP) via the sulfhydryl terminus of a phosphopantetheine group, which, in turn, is attached to a serine residue of the acyl carrier protein. ACP is a single polypeptide chain of 77 residues and can be regarded as a giant prosthetic group, a "macro CoA.". In mammals the generation of acetyl-ACP and malonyl-ACP are catalysed by a single bifunctional protein domain, malonyl-acetyl transferase (MAT).

The subsequent pathway of fatty acid synthesis from acetyl-ACP and malonyl-ACP involves the repetition of a 4-step reaction sequence: condensation, reduction, dehydration, and reduction.

Condensation - In the condensation reaction, a four-carbon unit acetoacetyl-ACP is formed from a two carbon unit acetyl-ACP and a three-carbon unit malonyl-ACP by the enzyme 3-ketoacyl-ACP synthetase (also called acyl-malonyl-ACP condensing enzyme) and CO_2 is released. The reason acetyl-ACP and malonyl-ACP are involved rather than two molecules of acetyl-ACP is that the equilibrium for the synthesis of acetoacetyl-ACP from two molecules of acetyl-ACP is highly unfavorable. In contrast, the equilibrium is favorable if malonyl-ACP is a reactant because its decarboxylation contributes a substantial decrease in free energy. In effect ATP, drives the condensation reaction even though it does not directly participate. Rather, ATP is used in the earlier step of carboxylation of acetyl-CoA to malonyl-CoA. It is the free energy thus stored in malonyl-CoA that is released in the decarboxylation reaction accompanying the formation of acetoacetyl-ACP. Although HCO_3^- is required for fatty acid synthesis, its carbon atom does not appear in the product. Rather, all the carbon atoms of fatty acids containing an even number of carbon atoms are derived from acetyl-CoA.

- Reduction – In the next step, acetoacetyl-ACP is reduced to d-3-hydroxybutyryl-ACP by the enzyme 3-ketoacyl-ACP reductase (or β-ketoacyl reductase) which reduces the carbon 3 ketone to a hydroxyl group. This reaction differs from the corresponding one in fatty acid degradation (discussed earlier) in two respects: (i) the 'd' rather than the 'l' isomer is formed; and (ii) NADPH is the reducing agent, whereas NAD^+ is the oxidizing agent in β oxidation. This difference exemplifies the general principle that NADPH is consumed in biosynthetic reactions, whereas NADH is generated in energy-yielding reactions.

- Dehydration and further reduction - The d-3-hydroxybutyryl-ACP is dehydrated (removes H_2O) by the enzyme 3-hydroxyacyl-ACP dehydratase to form crotonyl-ACP, which is a trans-Δ2-enoyl-ACP. The final step in the cycle, carried out by the enzyme enoyl-ACP reductase, reduces the C2-C3 double bond converting crotonyl-ACP to butyryl-ACP. NADPH is again the reductant, whereas FAD is the oxidant in the corresponding reaction in β-oxidation. These last three reactions - a reduction, a dehydration, and a second reduction - convert acetoacetyl-ACP into butyryl-ACP, which completes the first elongation cycle. Thus the first round of fatty acid synthesis from acetylCoA and malonylCoA results in the formation of a 4-carbon moiety.

In the second round of fatty acid synthesis, butyryl-ACP condenses with a new molecule of malonyl-ACP to form a C6-β-ketoacyl-ACP. Reduction, dehydration, and a second reduction convert the C6-β-ketoacyl-ACP into a C6-acyl-ACP (caproyl-ACP), which is ready for a third round of elongation. A further five successive rounds of synthesis, consuming additional molecules of malonyl-CoA, lead to the formation of 8-, 10-, 12-, 14- and 16-carbon moieties respectively, the elongation cycles continuing until the C16-palmitoyl-ACP is formed. This intermediate is a good substrate for the enzyme palmitoyl thioesterase which hydrolyzes C16- palmitoyl-ACP to yield palmitate and ACP. The thioesterase acts as a ruler to determine fatty acid chain length. Further elongation and the insertion of double bonds are carried out by other enzyme systems.

Overall the 7 cycles required for the synthesis of the C16 fatty acid palmitate require 8 molecules of acetyl-CoA (as one molecule of acetyl-CoA and 7 molecules of malonyl-CoA), 14 molecules of NADPH (given there are two reductive steps per cycle), and 7 molecules of ATP (required to generate the 7 molecules of malonyl-CoA from 7 molecules of acetyl-CoA). As discussed earlier, since one molecule of NADPH is generated for each molecule of acetyl-CoA that is formed by the action of ATP citrate lyase on each molecule of citrate that has moved from the mitochondrion to the cytosol, then eight molecules of NADPH will be formed when the eight molecules of acetyl-CoA required to form one molecule of palmitate are produced in this way. The additional six molecules of NADPH required for this process (14 required given 7 cycles with 2 reductive steps per cycle) come from the pentose phosphate pathway.

Further elongation and the insertion of double bonds are carried out by other enzyme systems and fatty acids with an odd number of carbon atoms are synthesized starting with propionyl-ACP rather than acetyl-ACP. Propionyl-ACP is formed from propionyl-CoA by the enzyme malonyl-acetyl transferase (MAT).

Organization of the Enzymes of Fatty Acid Synthesis in Mammals

The enzyme system in mammals that catalyzes the synthesis of saturated long-chain fatty acids from acetyl-CoA, malonyl-CoA, and NAPDH is called fatty acid synthase where the seven component enzymes involved are linked in a large polypeptide chain. Mammalian fatty acid synthase is a dimer of identical 260-kd subunits. Each chain is folded into three domains joined by flexible regions. Domain 1, the substrate entry and condensation unit, contains acetyl transacylase, malonyl transacylase, and 3-ketoacyl-ACP synthetase (condensing enzyme). Domain 2, the reduction unit, contains the acyl carrier protein (ACP), and the three enzymes 3-ketoacyl reductase, 3-hydroxyacyl-ACP dehydratase, and enoyl-ACP reductase. Domain 3, the palmitate release unit, contains the thioesterase. Thus, seven different catalytic sites are present on a single polypeptide chain improving coordination of the synthetic activity of different enzymes.

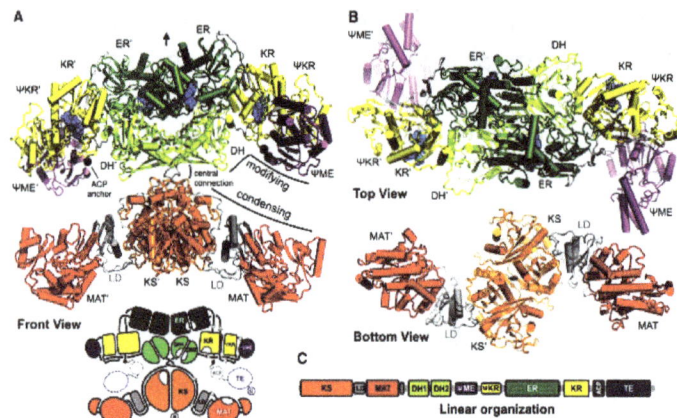

Figure: Structural overview of porcine fatty acid synthase.

(A) Side view cartoon representation of fatty acid synthase, colored by domains as indicated. Linkers and linker domains are depicted in gray. Bound NADP+ cofactors and the attachment sites for the disordered C-terminal ACP/TE domains are shown as blue and black spheres, respectively. The position of the pseudo-twofold dimer axis is depicted by an arrow at the top of the side view; domains of the second chain are indicated by an appended prime. The lower panel (front view) shows a corresponding schematic diagram. (B) Top (upper panel) and bottom (lower panel) views, demonstrating the "S" shape of the modifying (upper) and condensing (lower) parts of fatty acid synthase. The pseudo-twofold axis is indicated by an ellipsoid. (C) Linear sequence organization of fatty acid synthase, at approximate sequence scale. Note the C-terminal ACP and thioesterase domains of the fatty acid synthase polypeptide were not seen in the crystal structure presumably because of their flexibility. Protein flexibility may facilitate transfer of ACP-attached reaction intermediates among the several active sites in each half of the complex. Reproduced from Maier et al., 2008with permission. Click to enlargeThe crystal structure of porcine fatty acid synthase is shown in figure and reveals a complex architecture of alternating linkers and enzymatic domains. The enzyme assembles into an intertwined dimer approximating an "X" or back-to-back 'ƆC' shape. It is segregated into a lower condensing portion, containing the condensing β-ketoacyl synthase (KS) and the malonyl-acetyl transferase (MAT) domains, and an upper portion including the dehydratase (DH), NADPH-dependent enoyl reductase (ER), and NADPH−dependent β-ketoreductase (KR) domains responsible for β-carbon modification. The condensing and modifying parts of mammalian fatty acid synthase are loosely connected and form only tangential contacts. Substrate shuttling is facilitated by flexible tethering of the acyl carrier protein domain and by the limited contact between the condensing and modifying portions of the multienzyme, which are mainly connected by linkers rather than direct interaction. The structure identifies two additional nonenzymatic domains: (i) a pseudo-ketoreductase (ΨKR) and (ii) a peripheral pseudo-methyltransferase (ΨME) that is probably a remnant of an ancestral methyltransferase domain. The structural organization of domains deviates dramatically from their linear arrangement in sequence .

To summarise, in the priming step, the acetyl transferase loads acetyl-CoA onto the terminal thiol of the phosphopantetheine cofactor of the acyl carrier protein (ACP), which passes the acetyl moiety over to the active site cysteine of the β-ketoacyl synthase (KS). Malonyl transferase (MT) transfers the malonyl group of malonyl-CoA to ACP, and the KS catalyzes the decarboxylative condensation of the acetyl and malonyl moieties to an ACP-bound β-ketoacyl intermediate. The β-carbon position is then modified by sequential action of the NADPH–dependent β-ketoreductase (KR), a dehydratase (DH), and the NADPH-dependent enoyl reductase (ER) to yield a saturated acyl product elongated by two carbon units. This acyl group functions as a starter substrate for the next round of elongation, until the growing fatty acid chain reaches a length of 16 to 18 carbon atoms and is released from ACP. In mammalian fatty acid synthase, the malonyl and acetyl transferase reactions are catalyzed by a single bifunctional protein domain, the malonyl-acetyl transferase (MAT), and the products are released from ACP as free fatty acids by a thioesterase (TE) domain.

A multienzyme complex consisting of covalently joined enzymes is more stable than one formed by noncovalent attractions and intermediates can be efficiently handed from one active site to another without leaving the assembly. It seems likely that multifunctional enzymes such as fatty acid synthase arose in eukaryotic evolution by exon shuffling because each of the component enzymes is recognizably homologous to its bacterial counterpart.

Regulation of Fatty Acid Oxidation and Synthesis

Acetyl-CoA carboxylase - ACC is a central enzyme involved in fatty acid β-oxidation and fatty acid biosynthesis. ACC catalyzes the carboxylation of acetyl-CoA producing malonyl-CoA, which can be used by fatty acid synthase for fatty acid biosynthesis. While malonyl-CoA is used as a substrate for fatty acid biosynthesis, malonyl-CoA is also a potent inhibitor of mitochondrial fatty acid uptake secondary to inhibition of CPT1. There are two forms of ACC, a 265 kDa ACC1 isoform, which is highly expressed in the liver and adipose tissue, and a 280 kDa ACC2 isoform which is more specific to highly metabolic organs such as skeletal muscle and the heart. AMPK plays a major role in ACC1 and ACC2 regulation by phosphorylating and inhibiting ACC activity. In situations of increased energy demand, AMPK is activated, where it then phosphorylates and inactivates both isoforms of ACC. The adipocyte hormones adiponectin and leptin exert their effects on food intake, energy homeostasis, stimulation of fatty acid oxidation and glucose uptake by stimulating the phosphorylation and activation of AMPK in skeletal muscle (and liver – adiponectin). The inhibition of ACC by AMPK leads to a reduction of the molecules involved in gluconeogenesis in the liver, and reduction of glucose levels in vivo. ACC2 inhibition can lead to an increase in fatty acid β-oxidation, while fatty acid biosynthesis decreases when ACC1 is inhibited.

Several transcriptional factors can regulate ACC gene expression, including sterol regulatory element binding protein (SREBP1a and SREBP1c) and carbohydrate response element binding protein (ChREBP). SREBP is regulated by insulin, which promotes the endoplasmic reticulum

SREBP1c to be cleaved and translocated to the nucleus, leading to stimulation of ACC expression. ChREBP expression can be induced by high glucose concentrations, resulting in the activated ChREBP promoting the expression of ACC1 and fatty acid synthase. Nuclear respiratory factor-1 (NRF-1) is a principal modulator of mitochondrial protein expression and mitochondrial biogenesis, both of which are important for higher mitochondrial fatty acid β-oxidation capacity.

Malonyl-CoA decarboxylase - MCD is the enzyme responsible for decarboxylation of malonyl-CoA to acetyl-CoA. Generally, the level of malonyl-CoA is decreased when MCD activity is increased, resulting in an elevated rate of fatty acid oxidation. It has been reported that protein kinases that phosphorylate and inhibit ACC might activate MCD. However, MCD appears to be primarily regulated by transcriptional means. Therefore, MCD and ACC appear to work in harmony to regulate the pool of malonyl-CoA that can inhibit CPT1.

Membrane transport proteins - Regulation can occur at the level of fatty acid entry in to the cell. AMPK, PKC, and PPARγ positively regulate the activity of the fatty acid translocase CD36/FATP.

Carnitine palmitoyltransferase 1 - The CPT isoform, CPT1, resides on the inner surface of the outer mitochondrial membrane, and is a major site of regulation of mitochondrial fatty acid uptake. CPT1 is potently inhibited by malonyl-CoA, the product of ACC that binds to the cytosolic side of CPT1. Mammals express three isoforms of CPT1, which are encoded by different genes - the liver isoform (CPT1α), the muscle isoform (CPT1β), and a third isoform of CPT1 (CPT1c), which is primarily expressed in the brain and testis. More specifically, the heart expresses two isoforms of CPT1, an 82 KDa (CPT1α) isoform and the predominant 88 KDa (CPT1β) isoform (that has the highest sensitivity to malonyl-CoA inhibition). Insulin and thyroid hormone can regulate the sensitivity of CPT1α in the liver; whereas the muscle isoform CPT1β is not affected. Levels of malonyl-CoA are inversely correlated with fatty acid β-oxidation rates and studies on ACC2 knockout mice suggest two separate cellular malonyl-CoA pools, malonyl-CoA produced by ACC1 (used mainly for lipogenesis), and a cytosolic pool of malonyl-CoA produced by ACC2 involved in the regulation of CPT1 and fatty acid β-oxidation.

Allosteric control of β-oxidation enzymes - The activity of the enzymes of fatty acid β-oxidation is affected by the level of the products of their reactions. Each of the β-oxidation enzymes are inhibited by the specific fatty acyl-CoA intermediate it produces. Interestingly, 3-ketoacyl-CoA can also inhibit enoyl-CoA hydratase and acyl-CoA dehydrogenase. β-oxidation can also be allosterically regulated by the ratio of NADH/NAD+ and acetyl-CoA/CoA. A rise in the NADH/NAD+ or acetyl-CoA/CoA ratios results in inhibition of fatty acid β-oxidation. Increases in the acetyl-CoA/ CoA ratio have specifically been shown to lead to feedback inhibition of ketoacyl-CoA thiolase.

Transcriptional and post-transcriptional regulation – The proteins involved in fatty acid β-oxidation are regulated by both transcriptional and post-transcriptional mechanisms. PGC-1α, a transcription factor co-regulator, and the transcription factor PPARα act in the nucleus to increase transcription of mitochondrial genes, fatty acid utilization genes, and other transcription factors.

There are a number of transcription factors that regulate the expression of these proteins. The peroxisome proliferator-activated receptors (PPARs) and a transcription factor co-activator PGC-1α are the most well known transcriptional regulators of fatty acid β-oxidation. Examples of proteins involved in fatty acid β-oxidation that are transcriptionally regulated by the PPARs include FATP, acyl-CoA synthetase (ACS), CD36/FAT, MCD, CPT1, long chain acyl-CoA dehydrogenase, and medium chain acyl-CoA dehydrogenase. Estrogen-related receptor α (ERRα) has also been implicated in the regulation of fatty acid β-oxidation, having been shown to also regulate transcription of the gene encoding medium chain acyl-CoA dehydrogenase. Ligands that bind to and modulate the activity of PPARα, δ, and γ include fatty acids.

The transcriptional co-activator PGC-1α binds to and increases the activity of PPARs and ERRα to regulate fatty acid β-oxidation. PGC-1α modulates the activity of a number of transcription factors that can increase the expression of proteins involved in fatty acid β-oxidation, the TCA cycle, and the electron transport chain. Increasing PGC-1α protein expression induces massive mitochondrial biogenesis in skeletal muscle.

PGC-1α is regulated at both the gene and protein level. AMPK increases the activity of pre-existing PGC-1α protein and increases PGC-1α mRNA levels by regulating the binding of transcription factors to specific sequences located in the PGC-1α gene promoter. Free fatty acids can also regulate PGC-1α protein expression as a high fat diet can elevate levels of PGC-1α in rat skeletal muscle.

Lipoproteins

Lipoproteins are complex aggregates ('particles') of lipids and proteins that render the hydrophobic lipids compatible with the aqueous environment of body fluids and enable their transport throughout the body of all vertebrates and insects to tissues where they are required. Because of their clinical importance, a very high proportion of research on lipoproteins deals with their functions in humans in relation to health, and the discussion that follows reflects this. Lipoproteins are synthesised mainly in the intestines and liver. Within the circulation, these aggregates are in a state of constant flux, changing in composition and physical structure as the peripheral tissues take up the various components before the remnants return to the liver. The most abundant lipid constituents are triacylglycerols, free cholesterol, cholesterol esters and phospholipids (phosphatidylcholine and sphingomyelin especially), although fat-soluble vitamins and antioxidants are also transported in this way. Free (unesterified) fatty acids and lysophosphatidylcholine are bound to the protein albumin by hydrophobic forces in plasma and in effect are detoxified.

The circulating lipoproteins are structurally and metabolically distinct from the proteolipids containing covalently linked fatty acids or other lipid moieties.

Composition and Structure

Lipoprotein classes: Ideally, the lipoprotein aggregates should be described in terms of the different protein components known as apoproteins (or 'apolipoproteins'), as these determine the overall structures and metabolism, and the interactions with receptor molecules in liver and peripheral tissues. However, the practical methods that have been used to segregate different lipoprotein classes have determined the nomenclature. Thus, the main groups are classified as chylomicrons (CM), very-low-density lipoproteins (VLDL), low-density lipoproteins (LDL) and high-density lipoproteins (HDL), which are based on the relative densities of the aggregates on ultracentrifugation and with fortuitously broadly distinct functions. However, these classes can be further refined by improved separation procedures, and intermediate-density lipoproteins (IDL) and subdivisions of the HDL (e.g. HDL_1, HDL_2, HDL_3 and so forth) are often defined, and each of these may have distinctive apoprotein compositions and biological properties that for example can be relevant to cardiovascular disease. Density is determined largely by the relative concentrations of triacylglycerols (lighter) and proteins and by the diameters of the broadly spherical particles, which vary from about 6000Å in CM to 100Å or less in the smallest HDL.

Some compositional details are listed in table:

Table: Physical properties and lipid compositions of lipoprotein classes.				
	CM	VLDL	LDL	HDL
Density (g/ml)	< 0.94	0.94-1.006	1.006-1.063	1.063-1.210
Diameter (Å)	6000-2000	600	250	70-120
Total lipid (wt%) *	99	91	80	44
Triacylglycerols	85	55	10	6
Cholesterol esters	3	18	50	40
Cholesterol	2	7	11	7
Phospholipids	8	20	29	46
Most of the remaining material comprises the various apoproteins.				

The data for the relative compositions of the various lipid components should not be considered as absolute, as they are in a state of constant flux, but in general the lower the density class, the higher the proportion of triacylglycerols and the lower the proportions of phospholipids and the other lipid classes. In fact, the VLDL and LDL exhibit a continuum of decreasing size and density.

The fatty acid compositions of the main lipid classes in human lipoproteins are listed in table. As might be expected, the triacylglycerols tend to contain a high proportion of saturated and monoenoic fatty acids, while the phospholipids contain the highest proportion of polyunsaturated, especially arachidonate. Cholesterol esters are enriched in linoleate, reflecting their biosynthetic origin. The composition of the triacylglycerols of the chylomicrons (not listed) depends largely on that of the dietary fatty acids. In general, minor differences only occur for the compositions of each lipid among the lipoprotein classes.

Table: Fatty acid compositions (wt% of the total) in the main lipids of human lipoprotein classes.									
	Triacylglycerols			Cholesterol esters			Phospholipids		
	VLDL	LDL	HDL	VLDL	LDL	HDL	VLDL	LDL	HDL
16:0	27	23	23	12	11	11	34	36	32
18:0	3	3	4	1	1	1	15	14	14
18:1	45	47	44	26	22	22	12	12	12
18:2	16	16	16	52	60	55	20	19	21
20:4(n-6)	2	5	8	6	7	6	14	13	16
From Skipski, V.P. In: Blood Lipids and Lipoproteins. Quantitation, Composition and Metabolism.									

Apoproteins - Although a wide variety of proteins of various kinds are transported in the form of lipoprotein complexes, the apoproteins are the defining components that are essential for their formation and subsequent metabolism. In general, these consist of a single polypeptide chain often with relatively little tertiary structure, and they are required to solubilize the non-polar lipids in the circulation and to recognize specific receptors, which direct their metabolism and that of the associated lipoproteins. The various types with their main (but not exclusive) lipoprotein associations, molecular weights and broad functions are listed in table.

Table: The main properties of the apoproteins.			
Apoprotein	Molecular weight	Lipoprotein	Function
Apo A1	28,100	HDL, CM	Main structural protein. Lecithin:cholesterol acyltransferase (LCAT) activation.
Apo A2	17,400	HDL, CM	Enhances hepatic lipase activity
Apo A4	46,000	HDL, CM	May increase triacylglycerol secretion
Apo A5	39,000	HDL, VLDL, CM	Enhances triacylglycerol uptake
Apo B48	241,000	CM	Derived from Apo B100 – lacks the LDL receptor
Apo B100	512,000	IDL, LDL, VLDL	Binds to LDL receptor
Apo C1	7,600	HDL, VLDL, CM	Activates LCAT
Apo C2	8,900	HDL, VLDL, CM	Activates lipoprotein lipase
Apo C3	8,700	HDL, VLDL, CM	Inhibits lipoprotein lipase and controls triacylglycerol turnover
Apo D	33,000	HDL	Associated with LCAT, progesterone binding
Apo E	34,000	HDL, VLDL, CM	At least 3 forms. Binds to LDL receptor
Apo(a)	300,000-800,000	Lp(a)	Linked by disulfide bond to apo B100 and similar to plasminogen
Apo H, J, L			Poorly defined functions
Apo M		HDL	Transports sphingosine-1-phosphate
Roman numerals are often used to designate apoproteins (e.g. Apo AI, AII, AIII, etc) as an alternative. CM = chylomicrons			

Apo B100 and apo B48 large and water-insoluble and they are the only non-exchangeable apoproteins, which are assembled into triacylglycerol-rich lipoproteins with their lipid components in the intestines or liver. In humans, apo B48 is produced only in the small intestine in response to fat in the diet. Cholesterol esters are required for proper folding of apo B. With 4536 amino acid residues, apo B100 is one of the largest monomeric proteins known; apo B48 represents the N-terminal 48% of apo B100. These apoproteins stay with their lipid aggregates during their passage in plasma and the various metabolic changes that occur, until they are removed eventually via specific receptors.

The remaining soluble or exchangeable apoproteins, such as apo E, apo A4, apo C3, apo A5, and apo A1, are much smaller in molecular weight and can exchange between lipoprotein classes and acquire lipids during circulation.

Apo A1 is the main protein component of HDL, and is synthesised within the liver (70%) and intestine (30%). It is a 28-kDa single polypeptide consisting of 243 amino acids, which has no disulfide linkages or glycosylation. Apart from the 44 amino acid N-terminal region, the protein is arranged as eight α-helical segments of 22 amino acids with two 11-mer repeats, and in some instances these are separated by proline residues. It is believed that the helices are amphipathic, each with a hydrophobic face that interacts with lipids and a polar face that interacts with the aqueous phase. The molecule probably exists in several conformational forms, and because of this conformational adaptability, it can stabilize all HDL subclasses.

Apo A2 is the second most abundant HDL apoprotein, and it exists as a homodimer with two polypeptide chains, each 77 amino acids in length and linked by a disulfide bond. Apo A4 is the largest member of the exchangeable apoprotein family and is a 376-amino acid glycoprotein, which is synthesised in intestinal enterocytes and secreted as a constituent of chylomicrons. It transfers to HDL in plasma, but a high proportion is lipid free. Apo A5 is expressed in the liver mainly and although its concentration in plasma is relatively low, it is recognized as a potent regulator of triacylglycerol metabolism that leads to the enhanced lipolysis and the clearance of the lipoprotein remnants.

Apo C apoproteins are found in all lipoprotein classes. Apo C1 is the smallest exchangeable protein in this group and is synthesised in the liver, where like apo C3, the most abundant of the group, it tends to inhibit lipase activity. In contrast, apo C2 is especially important as an activator of lipoprotein lipase.

Apo D is atypical in that it is very different in structure from other apoproteins, and it is expressed widely in mammalian tissues (most others are produced mainly in liver and intestine). In plasma, it is present mainly in HDL and to a lesser extent in LDL, where it may function as a multi-ligand binding protein capable of transporting small hydrophobic molecules such as arachidonic acid, steroid hormones, and cholesterol for metabolism or signalling.

Apo E is an *O*-linked glycoprotein in three isoforms and is synthesised by many tissues, including liver, brain, adipose tissue, and arterial wall, but most is present in plasma lipoproteins derived primarily from the liver. It is involved in many aspects of lipid and lipoprotein homeostasis, both for the triacylglycerol-rich lipoproteins and HDL, and it is believed to have some non-lipid related functions, for example on immune response and inflammation. In addition, as lipoproteins cannot cross the blood-brain barrier, apo E is synthesised in the brain where it is the main apoprotein and is of particular importance for cholesterol metabolism.

Lipoprotein structures - Lipoproteins are spherical (VLDL, LDL, HDL) to discoidal (nascent HDL) in shape with a core of non-polar lipids, triacylglycerols and cholesterol esters, and a surface monolayer, ~20Å thick, consisting of apoproteins, phospholipids and non-esterified cholesterol, which serves to obscure the hydrophobic lipids and present a hydrophilic face to the aqueous phase as illustrated schematically for a triacylglycerol-rich chylomicron below.

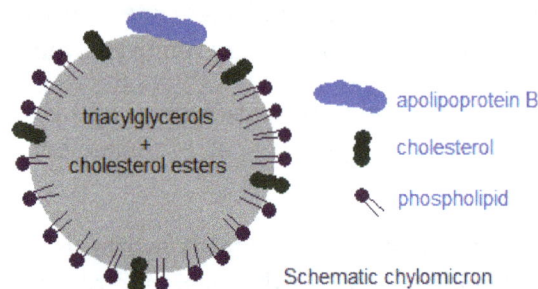

Schematic chylomicron

The physical properties of apoproteins enable them to bind readily at the interface between water and phospholipids, and specifically they bind to the phospholipids on the surface of the lipoproteins. In effect, this outer shell of amphipathic lipids and proteins solubilizes the hydrophobic lipid core in the aqueous environment. Each apoprotein, other than apo B100, tends to have a helical shape with a hydrophobic domain on one side that binds to the lipid core and a hydrophilic face that orientates to the aqueous phase. As the lipid compositions of the lipoproteins change during circulation throughout the body, the apoproteins are able to adapt to the altering affinities at the surface by changing conformation. For example, some have very little tertiary structure so are flexible, while apo A1 has a mobile or hinge domain. The polar nature of the surface monolayer prevents the lipoprotein particles from aggregating to form larger units. In addition, apoproteins have many different functions, some of which are listed in table. For example, some are ligands for receptors on cell surfaces and specify the tissues to which the lipid components are delivered, while others are cofactors for lipases or regulate lipid metabolism in the plasma in various ways.

LDL particles, for example, average 22 nm in diameter with roughly 3000 lipid molecules in total, and they contain a hydrophobic core of approximately 170 triacylglycerol, 1600 cholesterol ester and 200 unesterified cholesterol molecules. The amphipathic surface monolayer has a single copy of apo B100 together with about 700 phospholipid and 400 free cholesterol molecules. Phosphatidylcholine, about 450 molecules, and sphingomyelin, about 185 molecules, are the main phospholipids, together with smaller numbers of lysophosphatidylcholine, phosphatidylethanolamine and other lipid molecules. The structure and physical functions of LDLs depend mainly on the core–lipid composition and the conformation of the apoB-100, which is able to interact with extracellular membranes such as blood vessel intima where the LDL lipids are susceptible to modification, e.g. by acetylation, enzymatic digestion and oxidation.

In contrast, HDL are highly heterogeneous in terms of their size, lipid and protein contents, and their functional properties, and they can be separated by various means into subclasses, HDL_1, HDL_2, HDL_3, etc, that reflect the differences in composition. Discoidal nascent HDL particles are believed to consist of a small unilamellar bilayer, containing approximately 160 molecules of phospholipid, which is surrounded by four apoprotein molecules, including at least two apo A1 monomers. Although most HDL particles in human plasma are spherical, the structures are poorly characterized in comparison to discoidal HDL. It is believed that the apoA-I molecule changes conformation from the discoidal state and adopts a helical structure with the C-terminal domain binding to the phospholipids.

Lipoproteins can be categorized simplistically according to their main metabolic functions. For example, the principal role of the chylomicrons and VLDL is to transport triacylglycerols 'forward' as a source of fatty acids from the intestines or liver to the peripheral tissues. In contrast, the HDL remove excess cholesterol from peripheral tissues and deliver it to the liver where some is excreted in bile in the form of bile acids ('reverse cholesterol transport'). While these functions are considered separately for convenience in the discussion that follows, it should be recognized that the processes are highly complex and inter-related, and they involve transfer of apoproteins, enzymes and lipid constituents among the heterogeneous mix of all the lipoprotein fractions. In addition to the apoproteins, lipoproteins carry a number of enzymes with important functions, including lipases, acyltransferases, transport proteins and some with anti-oxidative or anti-inflammatory properties; some are concerned with metabolic processes that do not involve lipids.

As birds, amphibians, fishes and even round worms have lipoprotein systems comparable to those in mammals, it is evident that these must have developed early in evolution. The equivalent protein in these species is vitellogin, which is closely related to mammalian apo B.

Lipoprotein(a) (Lp(a)) is structurally and metabolically distinct from the other lipoproteins, and it consists of an LDL-like particle containing a specific highly polymorphic glycoprotein named apolipoprotein(a) (apo(a)), which is covalently bound via a disulfide bond to the apo B100 of the LDL-like particle. While its physiological function is uncertain, it is of particular interest in that there is an appreciable homology between apo(a) and the fibrinolytic proenzyme plasminogen and because it is a risk factor for cardiovascular disease. Apo(a) should not be confused with apo A.

Lipoprotein and Triacylglycerol Metabolism

Triacylglycerols are the most energy-dense molecules available to the body as a source of fuel but are highly hydrophobic. For efficient transport from the intestine and the liver to other organs of the body, it is essential that they be packaged in a form compatible with the aqueous environment in plasma, i.e. in lipoproteins. Chylomicrons and VLDL are mainly involved, although some proteins that are shared with HDL are essential for the process to function normally. For example, exchangeable apoproteins protect triacylglycerol-rich particles from non-specific interactions in plasma and ensure that they have the correct configuration to be acted upon by lipases.

Chylomicron formation: Dietary fatty acids and monoacylglycerols are absorbed by the enterocytes in the intestines, where they must cross the cytoplasm to the endoplasmic reticulum with the aid of fatty acid binding proteins. These are immediately utilized to form new triacylglycerols, and are thus detoxified, mainly by the monoacylglycerol pathway. The triacylglycerols are incorporated together with dietary cholesterol, much of which is in cholesterol ester form, into spherical chylomicron particles. These have a surface layer of phospholipids to which is attached a single molecule of the truncated form of apo B, apo B48, which is diagnostic for triacylglycerol-rich lipoproteins of intestinal origin. Chylomicrons are the largest lipoproteins present in the circulation, with their size dependent on the fed/fasted state, the rate of absorption of fat and the type and amount of fat absorbed.

The synthesis of apo B100 and its truncated form, and the accumulation of lipids to form chylomicrons or VLDL in intestinal cells and liver, respectively, are complex processes that are still only partly understood. Simplistically, secretory proteins such as apo B are synthesised on ribosomes on the surface of the endoplasmic reticulum and translocated through the membrane to the lumen of the endoplasmic reticulum. VLDL are then assembled by accretion of lipids, for example with the aid of a microsomal triacylglycerol transfer protein (MTTP), an essential protein that transfers phospholipids and triacylglycerols to nascent apo B for the assembly of lipoproteins. This occurs in three stages - pre-VLDL (pre-chylomicrons - nascent lipoproteins), VLDL2, a triacylglycerol-poor form of VLDL that is assembled in the Golgi and is transported to the basolateral membrane, where the final triacylglycerol-rich VLDL1 or chylomicrons with the assistance of apo B48 and apo A4, are secreted by a process of reverse exocytosis into the intestinal lamina propria. Apo A1 is generated separately in the endoplasmic reticulum of enterocytes, and it is transported to the Golgi and added to the chylomicrons just before the mature particle is secreted into the lymph.

The chylomicrons are transported via the intestinal lymphatic system and enter the blood stream at the left subclavian vein. During circulation throughout the body, triacylglycerols are removed by the peripheral tissues by endothelial-bound lipoprotein lipase with entry of fatty acids into muscle for energy production and adipocytes for storage. However, the apo B48 remains with the residual particle. The chylomicrons also contain some apo A1, which is synthesised in the intestines and liver, but this is transferred spontaneously to the HDL as soon as the chylomicrons reach the circulation, while transfer of apo E and apo C(1-3) in the reverse direction from the HDL to the surface of the chylomicrons, displacing apo A4, occurs at the same time. The depleted or 'remnant' chylomicrons, containing the dietary cholesterol, apo E and apo B48 mainly, eventually reach the liver where they are cleared from the circulation by a receptor-mediated process that requires the presence of apo E.

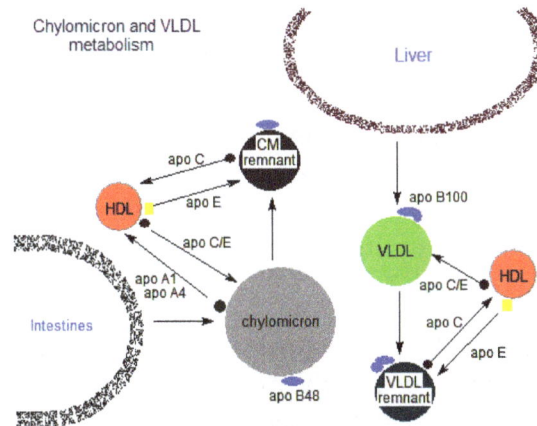

Liver catabolism - A high proportion of the VLDL remnants (or 'IDL') with apo B100 and apo E as the remaining proteins are sequestered in the liver perisinusoidal space (space of Disse) where they may undergo additional processing by lipases with further loss of triacylglycerols as they are converted to LDL. Both apoproteins are required for recognition of the VLDL remnants and LDL by the LDL receptors in the liver mainly, although many other tissues also contain analogous receptors. The main LDL receptor in liver is a polypeptide of 839 amino acids to which complex carbohydrate moieties are linked that spans the plasma membrane and has an extracellular domain, which is responsible for binding to apo B100 and apo E. Within the cell, the receptors cluster into regions of the plasma membrane known as 'coated pits', where the cytoplasmic leaflet is coated with the protein clathrin. After binding of the LDL and some of the VLDL remnants to the receptor, the LDL-receptor complexes are internalized by endocytosis of the coated pit and then dissociated by means of an ATP-dependent proton pump, which lowers the pH in the endosomes, enabling the receptors to be recycled to the plasma membrane. The LDL-containing endosomes fuse with lysosomes, and lipolytic enzymes, especially a lysosomal acid lipase (LAL), release free fatty acids and cholesterol from triacylglycerols and cholesterol esters, while acid hydrolases degrade the apoproteins. However, much of the apo E is believed to escape this process and is returned to the circulation and the HDL. An additional receptor, the LDL-receptor-related-protein, assists in the removal of chylomicron remnants.

After their release from lysosomes, the fatty acids and other lipid components serve as precursors for the synthesis of new lipid species and may also function in the regulation of many metabolic processes. For example, unesterified fatty acids are able to interact with the peroxisome proliferator-activated receptor PPARα and so target gene expression.

Secretion from the liver: The triacylglycerols of the remnant chylomicrons, together with cholesterol and cholesterol esters, are secreted by the liver into the circulation in the form of VLDL, which contain one molecule of the full-length form of apo B, apo B100. In addition, an appreciable amount of triacylglycerol in VLDL is synthesised in the liver from free fatty acids reaching it from adipose tissue via the plasma in the post-absorptive and fasted states, and stored in triacylglycerol form in lipid droplets for mobilization upon demand. In effect, liver lipid droplets and VLDL serve to buffer the plasma free fatty acids released following lipolysis in adipose tissue in excess of the requirements of muscle and liver.

Within the liver, the nascent VLDLs consisting largely of triacylglycerol droplets with the apo B100 are synthesised in the endoplasmic reticulum, and they are transported to the Golgi in a complex multistep process, involving a specific VLDL transport vesicle. In the lumen of the *cis*-Golgi lumen, VLDLs undergo a number of essential modifications before they are transported to the plasma membrane and secreted into the circulatory system. The surface layer of the newly synthesised VLDL is enriched in phosphatidylethanolamine, which rapidly exchanges with the phosphatidyl-choline of other lipoproteins. The newly synthesised VLDL contain a little apo C3, apo E and apo A5, which may have a role in the assembly process, but they rapidly take up apo C2 (10-20 mole-cules) and apo E from HDL after a few minutes in the circulation while the small amount of apo A1 of intestinal origin is transferred to HDL.

Lipoprotein lipase, the key enzyme in the peripheral tissues that is responsible for the hydrolysis of triacylglycerols from the chylomicrons and VLDL, is bound to the vascular surface of the endo-thelial cells of the capillaries of adipose tissue, heart and skeletal muscle, and lactating mammary gland primarily. It is a member of a lipase family that includes pancreatic lipase and hepatic tria-cylglycerol lipase, with an amino-terminal α/β-hydrolase domain harboring a catalytic triad and a carboxyl-terminal β-barrel domain that interacts with its substrate. The enzyme is synthesised in the endoplasmic reticulum where it is activated by lipase maturation factor 1 (LMF1), before the complex is stabilized with other chaperones so that it attains a proper tertiary fold for transport to the luminal surface of endothelial cells and into the interstitial space in the form of a monomer (not as a homodimer as was once believed). A small glycosylphosphatidylinositol-anchored pro-tein designated GPIHBP1 facilitates the transfer of lipoprotein lipase across the cell, and in con-cert with heparin sulfate-proteoglycans (HSPG) on the capillary wall anchors the enzyme to the endothelial cell surface. There, GPIHBP1 binds the enzyme in an appropriate conformation in a 1:1 complex mediated via the carboxyl-terminal domain to enable hydrolysis of the triacylglycerols of chylomicrons and LDL. Apo C2 is an absolute requirement for activation of the enzyme, and there is evidence that this opens a lid-like region of the enzyme to enable the active site to hydrolyse the fatty acid ester bonds of the triacylglycerols; apo A5 is also stimulatory. During lipolysis, several molecules of the enzyme, each activated by one molecule of apo C2, become attached to the surface of the chylomicron/VLDL particles simultaneously.

Lipoprotein lipase hydrolyses the primary ester linkages (positions *sn*-1/3) in triacylglycerols to produce free fatty acids and 2-monoacylglycerols. Then, the latter isomerize spontaneously to form 1/3-monoacylglycerols, which can be hydrolysed also by the enzyme. However, monoacyl-glycerols can be taken up directly by cells and are not found in the remnant lipoproteins or bound to circulating albumin. As the transport of VLDL particles progresses, the core of triacylglycerols is reduced and the proteins, including apo C2, and phospholipids on the surface are transferred away to the HDL. However, sufficient apo C2 remains to ensure that most of the triacylglycerols are removed. As partially delipidated lipoproteins are detected in the circulation, it is believed that there is a process of dissociation and rebinding to the enzyme, during each step of which triacylglycerols are hydrolysed and apo C2 is gradually released with formation of remnant particles. Lipoprotein lipase is also involved in the non-hydrolytic uptake of esters of cholesterol and retinol, possibly by facilitating transport.

Some of the unesterified fatty acids resulting from the action of lipoprotein lipase are taken up immediately by the cells by both receptor-mediated and receptor-independent pathways, where they can be used for energy purposes or for the synthesis of other lipids. The remainder is bound to circulating albumin from which it is released slowly to meet the cellular requirements of peripheral tissues. The glycerol produced is transported back to the liver and kidneys, where it can be converted to the glycolytic intermediate dihydroxyacetone phosphate. In muscle tissue, much of the fatty acids taken up are oxidized to two-carbon units, but in adipose tissue triacylglycerols are formed for storage purposes while in lactating mammary gland they are used for milk fat synthesis. During fasting, hormone-sensitive lipase releases fatty acids from the triacylglycerols stores and they are transported back into the circulation. Some lipoprotein lipase is present in the circulation, where it can continue to degrade the chylomicrons/VLDL.

Apo C1 and apo C3 inhibit lipoprotein lipase by competing for binding to lipoproteins rather than by deactivating the enzyme. Apo C3 inhibits the hepatic uptake of VLDL remnants also and so has a controlling influence on the turnover of triacylglycerols; high levels have been correlated with elevated levels of blood lipids (hypertriglyceridemia). In addition, angiopoietin-like proteins are key regulators of plasma lipid metabolism by serving as potent inhibitors of lipoprotein lipase. Improper regulation of the enzyme has been associated with the pathologies of atherosclerosis, coronary heart disease, cerebrovascular accidents, Alzheimer disease and chronic lymphocytic leukemia.

Lipid Metabolism

Lipid metabolism is a complex process that involves multiple steps involving the dietary intake of lipids (exogenous) or the production of lipids within the body (endogenous) to degradation or transformation (catabolism) into several lipid-containing structures in the body.

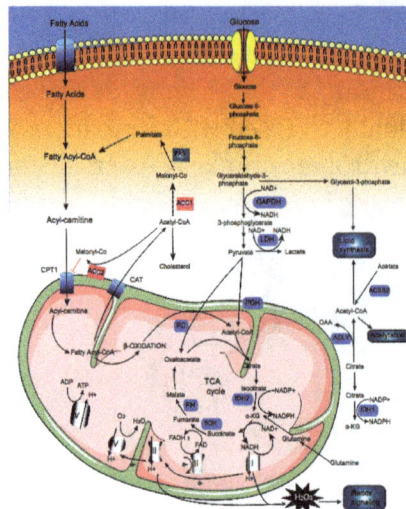

Figure: Lipid metabolism signaling pathway

Lipid metabolism mainly includes triglyceride (TG) metabolism, metabolism of cholesterol and its esters, and phospholipid and glycolipid metabolism. In these metabolic processes, many proteases, receptors, transcription factors, *etc.* are involved, and they are regulated by some signal transduction pathways, forming a complex and fine regulatory network to maintain the lipid metabolism balance of cells and the whole body. Lipid metabolism transduction signal pathways mainly include peroxisome proliferator-activated receptor (PPARs) signal transduction pathway, liver X receptor (LXRs) signal transduction pathway, sterol regulatory element binding protein (SREBPs) signal transduction guide route and so on. The lipid metabolism signal transduction pathways are complex, and there are many downstream target genes regulated by each pathway, and each pathway is also regulated by each other.

Lipid Metabolism Signaling Pathway

1. Lipid metabolism signaling pathway cascade: Lipid metabolism can be explained from catabolism and anabolism. In lipid catabolism reaction, triglyceride turns into glycerol fatty acid under the action of lipase; fatty acid, ATP, fatty acyl-CoA turns into fatty acyl-CoA, AMP, PPi under synthetase. Among them, lipase is hormone-sensitive triglyceride lipase, which is the rate-limiting enzyme for lipolysis. When the sympathetic nerve is excited, the secretion of adrenaline and norepinephrine increases, the receptor acting on the surface of the adipocyte membrane activates adenylate cyclase, promotes the synthesis of cyclized adenosine monophosphate, activates a protein kinase that is dependent on cyclic adenosine monophosphate, and phosphorylates triglyceride lipase in cytosol. Fatty acid production rate increases. Fatty acid CoA formed after fatty acid activation depends on the mitochondrial membrane on the mitochondrial transport mechanism into the mitochondrial inner membrane. Carnitine is the most effective transport factor for fat transport to mitochondrial oxidative energy supply. Appropriate supplementation of L-carnitine can increase the utilization of fatty acids. Fatty acid CoA enters the mitochondrial matrix and dehydrogenates from the beta carbon atom. Chemical dehydrogenation, thiolysis, a new generation of 2 carbon atoms of fatty acyl-CoA and 1 molecule of acetyl-CoA, is called β-oxidation of saturated fatty acids, and unsaturated fatty acids are also β-oxidized. Mammals fatty acids

in the body also have α-oxidation and ω-oxidation, and the product succinic acid enters the Krebs cycle to form CO_2, H_2O and a large amount of ATP. The 1 molecule of sugar is completely decomposed to form 36 molecules of ATP, and 1 molecule of soft fatty acid is completely decomposed to form 129 molecules ATP. Glycerol is converted into 3 molecules of glycerol phosphate by the action of glycerol phosphokinase, partially involved in the synthesis of triglycerides and phospholipids, partially into the glycolysis pathwayand part of the gluconeogenesis. Glycerol in a few tissues such as kidney and liver oxidation to CO_2 and H_2O, oxidation of one molecule of glycerol can produce 22 molecules of ATP. When glycerin is used for long-term endurance exercise, the utilization rate is increased, and it becomes an important substrate for gluconeogenesis. As an important source of blood sugar, it ensures relatively stable blood sugar in exercise. Excess carbohydrates in food are converted to triglycerides (TAG) in the liver, and very low-density lipoproteins (VLDLs). The process by which glucose is converted to fatty acids is called the de novo synthesis pathway (DNL) of fat, tightly regulated by hormones and nutritional status. In starvation, de novo synthesis of fat remains low level due to elevated blood glucagon and activation of the intracellular cyclic adenosine monophosphate (cAMP) pathway. After eating, elevated blood glucose and insulin levels stimulate the insulin signaling pathway, leading to protein kinases such as phosphatidylinositol 3-kinase (PI3K), AKT, atypical protein kinase C (aPKC), and mammalian rapamycin target protein complexes. Activation of lipids by mTORCs and protein phosphatases such as protein phosphatase 1 (pp1) and protein phosphatase 2 (pp2). A variety of proteases are involved in the regulation of fatty acid and triglyceride synthesis processes. These enzymes have very low activity when starved and are highly active after eating, thereby maintaining the balance of lipid metabolism in the body.

2. Pathway regulation Carnitine: The role of carnitine in fat metabolism has been confirmed, that is, triacylglycerol decomposes long-chain fatty acids of more than ten carbons. Oxidative energy supply must first be activated in the myocyte cytoplasm with coenzyme A (CoA) in the presence of ATP to form fatty acyl groups. CoA must be transferred to the mitochondria to oxidatively decompose, but the fatty acyl-CoA cannot pass through the mitochondrial inner membrane, and the carnitine can be used as a long-chain fatty acid carrier to oxidize and supply energy through the mitochondrial inner membrane, which acts as a 3-hydroxyl acceptor of carnitine. The acyl group of fatty acyl-CoA is transferred into the mitochondrial inner membrane, which is then catalyzed by carnitine lipid acyl-transferase II to separate the carnitine from the fatty acyl-CoA, and then reform the fatty acyl-CoA, which is continuously in the mitochondrial interstitial. β-oxidation, the resulting acetyl-CoA is decomposed into H_2O and CO_2 by the Krebs cycle, and at the same time releases a large amount of energy. The oxidation sites of lipids and acids are on the mitochondria, and the fatty acyl-CoA cannot enter the mitochondria, but the carnitine can carry the acyl group into the mitochondria to complete the fatty acid metabolism, so if the carnitine concentration is elevated, it can promote fatty acyl transport, thereby promoting fatty acid and fat metabolism. calmodulin, $1,25(OH)_2$-D_3 stimulates adipocyte Ca^{2+} influx to promote the activation of lipase synthase (FAS) and inhibit lipolysis, and to some extent, the dose-effect relationship is increased subsequently. An affinity membrane-bound protein, when specifically bound to $1,25(OH)_2$-D_3, mediates the regulation of intracellular Ca^{2+} concentration, which in turn affects lipid synthesis and decomposition. In the low-calcium diet, the level of $1,25(OH)_2$-D_3 in the body is automatically increased. By

binding to the membrane vitamin D receptor (mVDR), it stimulates a large amount of Ca^{2+} influx, eventually leading to an increase in triglycerides in the fat cells, the effect of dietary calcium on energy metabolism is related to the change of metabolic rate. the temperature of the body center of rats fed with high calcium diet and the expression of uncoupling protein UCP-2 is enhanced, and the energy utilization efficiency is decreased. UCP-2 is widely present in white adipose tissue. The main function is to control the production of heat, regulate insulin secretion and fatty acid utilization. It is observed that there is a direct dose-effect negative correlation between $1,25(OH)_2$-D_3 and UCP-2 expression, and this effect is not dependent. In addition, UCP-2 regulates the function of insulin secretion by $1,25(OH)_2$. The regulation of D_3 remains to be further studied. Once it is confirmed, it will provide a powerful theoretical supplement for the initial mechanism of calcium regulation of fat metabolism. Ca^{2+} is not only a key link in the regulation of energy metabolism by $1,25(OH)_2$-D_3. At the same time, it is also an important link in the regulation of fat metabolism by other factors. And *agouti* is an obese gene expressed in human fat cells, and its target product of the carboxy terminus of Agouti protein is calcium ion channel in human body. These phenomena can be mimicked by activating receptor-dependent and voltage-dependent calcium channels and can be reversed by inhibiting Ca^{2+} channels. Experiments have shown that Agouti-induced obese rats are treated with calcium channel antagonists for 4 weeks. Obesity has been significantly improved. The high expression of Agouti protein in obesity patients induces the increase of calcium influx in fat cells, which is also one of the causes of energy metabolism disorder.

Cholesterol

Cholesterol is a ubiquitous component of all animal tissues (and of some fungi), where much of it is located in the membranes, although it is not evenly distributed. The highest proportion of unesterified cholesterol is in the plasma membrane (roughly 30-50% of the lipid in the membrane or 60-80% of the cholesterol in the cell), while mitochondria and the endoplasmic reticulum have much less (~5% in the latter), and the Golgi contains an intermediate amount. Cholesterol is also enriched in early and recycling endosomes, but not in late endosomes. It may surprise some to learn that the brain contains more cholesterol than any other organ, where it comprises roughly a quarter of the total free cholesterol in the human body, 70-80% of which is in the myelin sheath. Of all the organic constituents of blood, only glucose is present in a higher molar concentration than cholesterol. In animal tissues, it occurs in the free form, esterified to long-chain fatty acids (cholesterol esters), and in other covalent and non-covalent linkages, including an association with the plasma lipoproteins. In plants, it tends to be a minor component only of a complex mixture of structurally related 'phytosterols', although there are exceptions, but it is nevertheless importance as a precursor of some plant hormones.

Animals in general synthesise a high proportion of their cholesterol requirement, but they can also ingest and absorb appreciable amounts from foods. On the other hand, many invertebrates, including insects, crustaceans and some molluscs cannot synthesise cholesterol and must receive it from the diet; for example, spiny lobsters must obtain exogenous cholesterol to produce essential sex hormones. Similarly, it must be supplied from exogenous sources to the primitive nematode

Caenorhabditis elegans, where it does not appear to have a major role in membrane structure, other than perhaps in the function of ion channels, although it is essential the production of steroidal hormones required for larval development; its uptake is regulated by the novel lipid phosphoethanolamine glucosylceramide. Some species are able to convert dietary plant sterols such as β-sitosterol to cholesterol. Prokaryotes lack cholesterol entirely with the exception of some pathogens that acquire it from eukaryotic hosts to ensure their intracellular survival; bacterial hopanoids are often considered to be sterol surrogates.

Cholesterol has vital structural roles in membranes and in lipid metabolism in general with an extraordinary diversity of biological roles, including cell signalling, morphogenesis, lipid digestion and absorption in the intestines, reproduction, stress responses, sodium and water balance, and calcium and phosphorus metabolism. It is a biosynthetic precursor of bile acids, vitamin D and steroid hormones (glucocorticoids, oestrogens, progesterones, androgens and aldosterone), and it is found in covalent linkage to specific membrane proteins or proteolipids ('hedgehog' proteins), which have vital functions in embryonic development. In addition, it contributes substantially to the development and working of the central nervous system. On the other hand, excess cholesterol in cells can be toxic, and a complex web of enzymes is essential to maintain the optimum concentrations. Because plasma cholesterol levels can be a major contributory factor to atherogenesis, media coverage has created what has been termed a 'cholesterophobia' in the population at large.

One of the main function of cholesterol is to modulate the fluidity of membranes by interacting with their complex lipid components, specifically the phospholipids such as phosphatidylcholine and sphingomyelin. As an amphiphilic molecule, cholesterol is able to intercalate between phospholipids in lipid bilayers to span about half a bilayer. In its three-dimensional structure, it is in essence a planar molecule that can interact on both sides. The tetracyclic ring structure is compact and very rigid. In addition, the location of the hydroxyl group facilitates the orientation of the molecule in a membrane bilayer, while the positions of the methyl groups appear to maximize interactions with other lipid constituents.

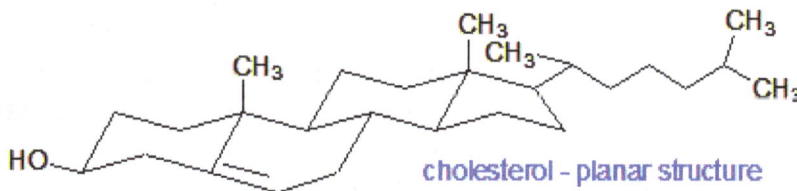

cholesterol - planar structure

As the α-face of the cholesterol nucleus (facing down) is 'smooth', it can make good contact with the saturated fatty-acyl chains of phospholipids down to about their tenth methylene group; the β-face (facing up) is made 'rough' by the projection of methyl groups from carbons 10 and 13. The interaction is mainly via van der Waals and hydrophobic forces with a contribution from hydrogen bonding of the cholesterol hydroxyl group to the polar head group and interfacial regions of the phospholipids, especially sphingomyelin. Intercalated cholesterol may also disrupt electrostatic interactions between the ionic phosphocholine head groups of nearby membrane phospholipids, leading to increased mobility of the head groups. Indeed, there is evidence that cholesterol forms stoichiometric complexes with the saturated fatty acyl groups of sphingomyelin and to a lesser extent of phosphatidylcholine.

Experiments with mutant cell lines and specific inhibitors of cholesterol biosynthesis suggest that an equatorial hydroxyl group at C-3 of sterols is essential for the growth of mammalian cells. The Δ5 double bond ensures that the molecule adopts a planar conformation, and this feature also appears to be essential for cell growth, as is the flexible *iso*-octyl side-chain. The C-18 methyl group is crucial for the proper orientation of the sterol. While plant sterols appear to be able to substitute for cholesterol in supporting many of the bulk properties of membranes in mammalian cells *in vitro*, cholesterol is essential for other purposes.

In the absence of cholesterol, a membrane composed of unsaturated lipids is in a fluid state that is characterized by a substantial degree of lipid chain disorder, i.e. it constitutes a *liquid-disordered* phase. The function of cholesterol is to increase the degree of order (cohesion and packing) in membranes, leading to formation of a *liquid-ordered* phase. In contrast, it renders bilayers composed of more saturated lipids, which would otherwise be in a solid gel state, more fluid. Thus, cholesterol is able to promote and stabilize a liquid-ordered phase over a substantial range of temperatures and sterol concentrations. Further, high cholesterol concentrations in membranes reduce their passive permeability to solutes. These effects permit the fine-tuning of membrane lipid composition, organization and function. In comparison to other lipids, cholesterol can flip rapidly between the leaflets in a bilayer, so the trans-bilayer distribution of cholesterol in some biological membranes is uncertain. While some models propose that cholesterol is in the outer leaflet, other studies suggest that most of the sterol is in the inner leaflet of human erythrocytes, for example, although in general the consensus at the moment is that cholesterol tends to be roughly equally distributed between the two leaflets of the plasma membrane. This distribution is important in that cholesterol promotes negative curvature of membranes and may be a significant factor in bringing about membrane fusion as in the process of exocytosis.

Cholesterol also has a key role in the lateral organization of membranes and their free volume distribution, factors permitting more intimate protein-cholesterol interactions that may regulate the activities of membrane proteins. Many membrane proteins bind strongly to cholesterol, including some that are involved in cellular cholesterol homeostasis or trafficking and contain a conserved region termed the 'sterol-sensing domain'. Some proteins bind to cholesterol deep within the hydrophobic core of the membrane via binding sites on the membrane-spanning surfaces or in cavities or pores in the proteins, driven by hydrogen bond formation. For example, cholesterol has an intimate interaction with G-protein-coupled receptors (GPCRs), and it is essential for the stability and function of the β2-adrenergic receptor and of ion pumps such as the (Na^+-K^+)-ATPase, which have specific binding site for cholesterol molecules. The last is the single most important consumer of ATP in cells and is responsible for the ion gradients across membranes that are essential for many cellular functions; depletion of cholesterol in the plasma membrane deactivates these ion pumps. The agonist affinities of other GPCRs, including the oxytocin and serotonin receptors, are dramatically increased by membrane cholesterol, while the inactive state of rhodopsin is stabilized through indirect effects on plasma membrane curvature. In the brain in addition to being essential for the structure of the myelin sheath, cholesterol is a major component of synaptic vesicles and controls their shape and functional properties, and it also has an important role in the organization and positioning of neurotransmitter receptors. In the nucleus of cells, cholesterol is intimately involved in chromatin structure and function.

Cholesterol forms a well-defined and essential association with the sphingolipids in the formation of the membrane sub-domains known as rafts, which are so important in the function of cells. It appears that the synthesis of cholesterol and of the sphingolipids, especially sphingomyelin, is regulated coordinately to satisfy the requirements of membrane composition and function. The interaction of cholesterol with ceramides is essential for the barrier function of the skin.

Simplistically, the higher cholesterol concentrations in the plasma membrane support its barrier function by increasing membrane thickness and reducing its permeability to small molecules. In contrast, the endoplasmic reticulum has increased membrane flexibility because of its lower cholesterol concentrations and thus enables the insertion and folding of proteins in its lipid bilayer.

Cholesterol Biosynthesis

Cholesterol biosynthesis involves a highly complex series of at least thirty different enzymatic reactions, which were unravelled in large measure by Konrad Bloch and Fyodor Lynen, When the various regulatory, transport and genetic studies of more recent years are taken into account, it is obvious that this is a subject that cannot be treated in depth here. In plants, cholesterol synthesis occurs by a somewhat different pathway with cycloartenol as the key intermediate.

Almost all nucleated cells are able to synthesise their full complement of cholesterol. The first steps involve the synthesis of the important intermediate mevalonic acid from acetyl-CoA and acetoacetyl-CoA, both of which are in fact derived from acetate, in two enzymatic steps. The acetyl-CoA precursor is in the cytosol as is the first enzyme, hydroxymethyl-glutaryl(HMG)-CoA synthase. The second enzyme HMG-CoA reductase is a particularly important control point, and is widely regarded as the rate-limiting step in the overall synthesis of sterols; its activity is regulated at the transcriptional level and by many more factors including a cycle of phosphorylation-dephosphorylation. This and subsequent enzymes are membrane-bound and are located in the endoplasmic reticulum. The enzyme HMG-CoA reductase is among the targets inhibited by the drugs known as 'statins', so that patients must then obtain much of their cholesterol from the diet via the circulation.

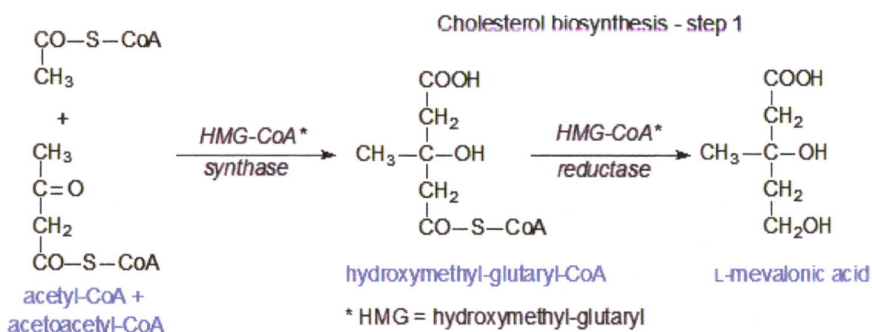

Cholesterol biosynthesis - step 1

The next sequence of reactions involves first the phosphorylation of mevalonic acid by a mevalonate kinase to form the 5monophosphate ester, followed by a further phosphorylation to yield an unstable pyrophosphate, which is rapidly decarboxylated to produce 5-isopentenyl pyrophosphoric acid, the universal isoprene unit. An isomerase converts part of the latter to 3,3-dimethylallyl pyrophosphoric acid.

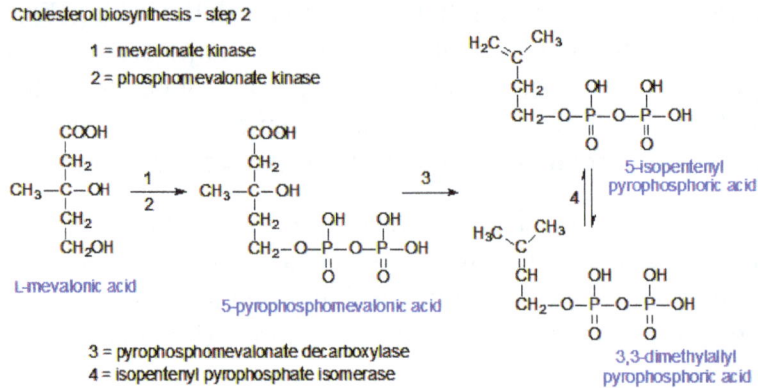

Cholesterol biosynthesis – step 2

1 = mevalonate kinase
2 = phosphomevalonate kinase

L-mevalonic acid

5-pyrophosphomevalonic acid

3 = pyrophosphomevalonate decarboxylase
4 = isopentenyl pyrophosphate isomerase

5-isopentenyl pyrophosphoric acid

3,3-dimethylallyl pyrophosphoric acid

5-Isopentenyl pyrophosphate is a nucleophile, but the isomerized product is electrophylic, facilitating the first step in the third series of reactions in which 5-isopentenyl pyrophosphate and 3,3-dimethylallyl pyrophosphate condense with the elimination of pyrophosphoric acid to form the monoterpenoid derivative geranyl pyrophosphate. This reacts with another molecule of 5-isopentenyl pyrophosphate to produce the sesquiterpene derivative (C_{15}) farnesyl pyrophosphate, two molecules of which are condensed to yield presqualene pyrophosphate. In turn, this is reduced by NADPH to produce a further key intermediate squalene. Both of the last steps are catalysed by the enzyme squalene synthase, which regulates the flow of metabolites into either the sterol or non-sterol pathways (with farnesyl pyrophosphate as the branch point) and is considered to be the first committed enzyme in cholesterol biosynthesis.

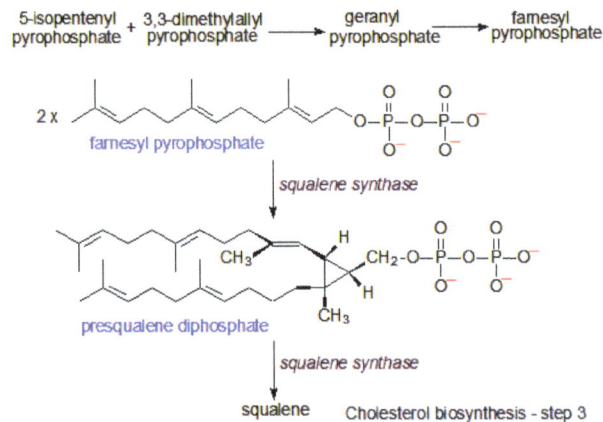

5-isopentenyl pyrophosphate + 3,3-dimethylallyl pyrophosphate → geranyl pyrophosphate → farnesyl pyrophosphate

2 x

farnesyl pyrophosphate

squalene synthase

presqualene diphosphate

squalene synthase

squalene   Cholesterol biosynthesis - step 3

In the next important step, squalene is first oxidized by a squalene monooxygenase to squalene 2,3-epoxide, which undergoes cyclization catalysed by the enzyme squalene epoxide lanosterol-cyclase to form the first steroidal intermediate lanosterol (or cycloartenol en route to phytosterols in photosynthetic organisms). In this remarkable reaction, there is a series of concerted 1,2-methyl group and hydride shifts along the chain of the squalene molecule to bring about the formation of the four rings. No intermediate compounds have been found. This is believed to be one of the most complex single enzymatic reactions ever to have been identified, although the enzyme involved is only 90 kDa in size. Again, the reaction takes place in the endoplasmic reticulum, but a cytosolic protein, sterol carrier protein 1, is required to bind squalene in an appropriate orientation in the presence of the cofactors NADPH, flavin adenine dinucleotide (FAD) and O_2; the reaction is promoted by the presence of phosphatidylserine.

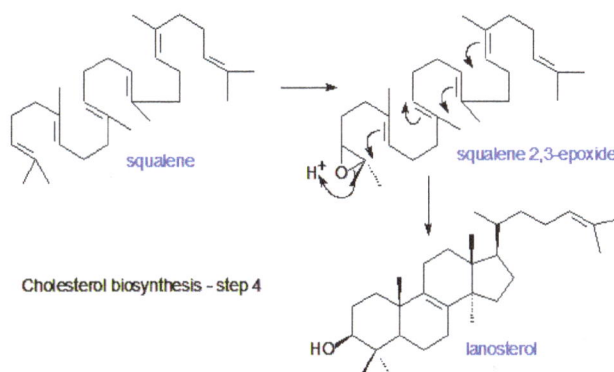

Cholesterol biosynthesis – step 4

In subsequent steps, lanosterol is converted to cholesterol by a series of demethylations, desaturations, isomerizations and reductions, involving 19 separate reactions. Thus, demethylation reactions produce zymosterol as an intermediate, and this is converted to cholesterol via a series of intermediates, all of which have been characterized, and by at least two pathways that utilize essentially the same enzymatic machinery but differ in the order of the various reactions, mainly at the point at which the Δ24 double bond is reduced. Desmosterol is the key intermediate in the co-called 'Bloch' pathway, while 7-dehydrocholesterol is the immediate precursor in the 'Kandutsch-Russel' pathway. While some tissues, such as adrenal glands and testis, use the Bloch pathway mainly, the brain synthesises much of its cholesterol by the 'Kandutsch-Russell' pathway. This may enable production of a variety of other minor sterols for specific biological purposes in different cell types/locations.

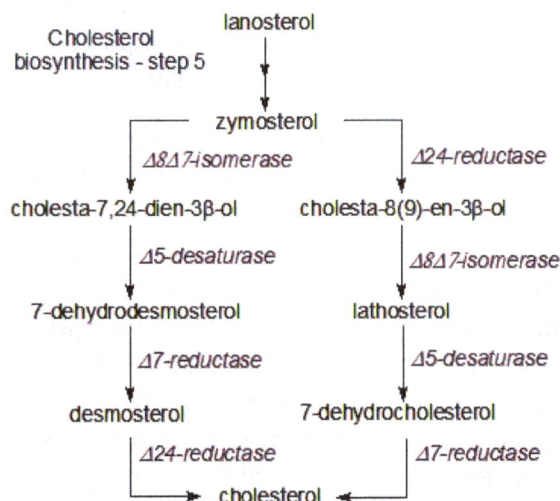

Cholesterol biosynthesis – step 5

Synthesis occurs mainly in the liver, although the brain, peripheral nervous system and skin synthesise their own considerable supplies. Cholesterol is exported from the liver and transported to other tissues in the form of low-density lipoproteins (LDL) for uptake via specific receptors. In animals, cells can obtain the cholesterol they require either from the diet via the circulating LDL, or they can synthesise it themselves as outlined above. Cholesterol biosynthesis is highly regulated with rates of synthesis varying over hundreds of fold depending on the availability of an external source of cholesterol, and cholesterol homeostasis requires the actions of a complex web of enzymes, transport proteins, and membrane-bound transcription factors.

Regulation of Cholesterol Homeostasis

In humans, only about a third of the body cholesterol is of dietary origin (mainly eggs and red meat), the remainder is produced by synthesis *de novo*. Although HMG-CoA reductase is recognized as the key enzyme in the regulation of cholesterol biosynthesis, the second rate-limiting enzyme in cholesterol biosynthesis is squalene monooxygenase, which undergoes cholesterol-dependent proteasomal degradation when cholesterol is in excess, guided by a 12 amino acid hydrophobic sequence on the enzyme that can serve as a degradation signal. When the cholesterol concentration in the endoplasmic reticulum is high, the degradation sequence detaches from the membrane and is exposed to provide the signal for the enzyme to be degraded.

Cholesterol in the endoplasmic reticulum is transferred to the Golgi and eventually to the plasma membrane by non-vesicular transport mechanisms involving in part soluble sterol transport proteins and partly by binding to those proteins that are intimately involved in the transport and metabolism of polyphosphoinositides such as phosphatidylinositol 4,5-bisphosphate (PI(4,5)P2). In the latter mechanism, cholesterol is transported by the monomeric form of oxysterol binding protein (OSBP)-related protein 2 (ORP2), which binds to PI(4,5)P2 in the plasma membrane to transfer its cargo. ORP2 then forms a tetramer that incorporates PI(4,5)P2 by binding at a C-terminal lipid transport domain (ORD) and removes it from the membrane for transport to other membranes such as the endosomal compartment. ORP2 is thus a key regulator of both cholesterol and PI(4,5)P2 concentrations in the plasma membrane. Any cholesterol in cellular membranes in excess of the stoichiometric requirement can escape back readily into the cell, where it may serve as a feedback signal to down-regulate cholesterol accumulation. Some of this 'active' cholesterol is converted to the relatively inert storage form, i.e. cholesterol esters, while some is used for steroidogenesis. In peripheral tissues, excess cholesterol is exported to high-density lipoproteins (HDL) and returned to the liver - reverse cholesterol transport.

The liver is the important for cholesterol synthesis, but it is essential for its elimination from the body in bile. In addition, the intestines play a major part in cholesterol homeostasis via absorption of dietary cholesterol and fecal excretion of cholesterol and its metabolites. A specific transporter in the brush border membrane of enterocytes in the proximal jejunum of the small intestine is involved in uptake of cholesterol from the intestinal contents, while the metabolism of sterols in the intestines is controlled mainly by an acetyl-CoA acetyltransferase (ACAT2), which facilitates intracellular cholesterol esterification, and the microsomal triglyceride transfer protein (MTTP), which is involved in the assembly of chylomicrons for export into lymph. There is evidence that dietary cholesterol or that synthesised *de novo* is necessary to maintain intestinal integrity, as cholesterol derived from circulating lipoproteins is not sufficient for the purpose.

In the intestines and especially the colon, the intestinal microflora are able to hydrogenate cholesterol from bile, diet and desquamated cells to form coprostanol with an efficiency that is dependent on the composition of microbial species. Coprastanol is not absorbed by the intestinal tissue to a significant extent, but it may inhibit the uptake of residual cholesterol.

Many other factors are involved in cholesterol homeostasis. For example, the regulatory element-binding proteins (mainly SREBP-1c and SREBP-2), which contain an N-terminal membrane domain and a C-terminal regulatory domain, are also essential to the maintenance of cholesterol homeostasis. Each is synthesised as an inactive precursor that is inserted into the endoplasmic reticulum where it can encounter an escort protein termed SREBP cleavage-activating protein (SCAP), which is the cellular cholesterol sensor. When the latter recognizes that cellular cholesterol levels are inadequate, it binds to the regulatory domain of SREBP. The SCAP-SREBP complex then moves to the Golgi, where two specific proteases (designated site-1 and site-2 proteases) cleave the SREBP enabling the C-terminal regulatory domain to enter the nucleus. There it activates transcription factors, such as the nuclear liver X receptor (LXR), which stimulate the expression of the genes coding for the LDL receptor in the plasma membrane and for the key enzyme in cholesterol biosynthesis, HMG-CoA reductase. This in turn stimulates the rate of cholesterol uptake and synthesis. Conversely, when cholesterol in the endoplasmic reticulum exceeds a threshold, it binds to SCAP in such a way that it prevents the SCAP-SREBP complex from leaving the membrane for the nucleus. Cholesterol synthesis and uptake are thereby repressed, and cholesterol homeostasis is restored. Then, the ATP binding cassette (ABC) transporters ABCA1 and ABCG1 in the plasma membrane, which contains much of the cellular cholesterol, are activated to export the excess. In addition, nuclear factor erythroid 2 related factor-1 or NRF1 in the endoplasmic reticulum binds directly to cholesterol and senses when its level is high to bring about a de-repression of genes involved in cholesterol removal, also with mediation by the liver X receptor.

Further regulation of cholesterol biosynthesis is exerted by sterol intermediates in cholesterol biosynthesis, such as lanosterol and 24,25-dehydrolanosterol, and by the side-chain oxysterols, especially 25-hydroxycholesterol, which can suppress the activation of SREBP by binding to an oxysterol-sensing protein in the endoplasmic reticulum or by direct effects on the biosynthetic and transport enzymes. However, many other factors are involved in maintaining within precise limits the large differences in cholesterol concentrations among the various membranes and organelles in cells. These include other regulatory proteins, and mechanisms that can involve either vesicle formation or non-vesicular pathways that utilize specific transport proteins, such as the ABC transporters. ceramide down-regulates cholesterol synthesis – another link between cholesterol and sphingolipid metabolism. There is also a link to phosphoinositide metabolism as phosphatidylinositol 4-phosphate has been described as a 'fuel' that drives sterol transport and allows the establishment of active sterol concentration gradients across membrane-bound compartments.

As cholesterol is unable to cross the blood brain barrier, all the cholesterol in brain is synthesised *de novo*, mainly in astrocytes, and it is transported to neurons in the form of Apo E complexes. In the brain and central nervous system, cholesterol synthesis is regulated independently of that in peripheral tissues, mainly by a form of the liver X receptor (LXR), while Apo E transcription is regulated by 24-hydroxy-cholesterol.

Cholesterol Catabolism

Cholesterol is not readily degraded in animal tissues so does not serve as a metabolic fuel. Only the liver possesses the enzymes to degrade significant amounts, and then via pathways that do not lead to energy production. Cholesterol and oxidized metabolites (oxysterols) are transferred back from peripheral tissues in lipoprotein complexes to the liver for catabolism by conversion to oxysterols and bile acids. The latter are exported into the intestines to aid digestion, while leading to some loss that is essential for cholesterol homeostasis. Until recently, it was believed that approximately 90% of cholesterol elimination from the body occurred via bile acids in humans. However, experiments with animal models now suggest that a significant amount is secreted directly into the intestines by a process known as trans-intestinal cholesterol efflux. How this occurs and its relevance to humans are under active investigation.

Certain bacterial species contain a 3β-hydroxysteroid:oxygen oxidoreductase (EC 1.1.3.6), commonly termed cholesterol oxidase, a flavoenzyme that catalyses the oxidation of cholesterol to cholest-5-en-3-one which is then rapidly isomerized to cholest-4-en-3-one as the first essential step in a more comprehensive catabolism of sterols. The enzyme is widespread in organisms that degrade organic wastes, but it also present in pathogenic organisms where it influences the virulence of infections. In biotechnology, it has been used for the production of a number of steroids, and it is employed in a clinical procedure for the determination of cholesterol levels in serum.

Cholesterol Esters

Cholesterol esters, i.e. with long-chain fatty acids linked to the hydroxyl group, are much less polar than free cholesterol and appear to be the preferred form for transport in plasma and as a biologically inert storage or de-toxification form to buffer an excess. They do not contribute to membrane structures but are packed into intracellular lipid droplets. Cholesterol esters are major constituents of the adrenal glands, and they accumulate in the fatty lesions of atherosclerotic plaques. Similarly, esters of steroidal hormones are also present in the adrenal glands, where they are concentrated in cytosolic lipid droplets adjacent to the endoplasmic reticulum; 17β-estradiol, the principal oestrogen in fertile women, is transported in lipoproteins in the form of a fatty acid ester.

Because of the mechanism of synthesis, plasma cholesterol esters tend to contain relatively high proportions of the polyunsaturated components typical of phosphatidylcholine. Arachidonic and "adrenic" (20:4(n-6)) acids can be especially abundant in cholesterol esters from the adrenal gland.

Table: Fatty acid composition of cholesterol esters (wt % of the total) from various tissues.							
	Fatty acids						
	16:0	18:0	18:1	18:2	18:3	20:4	22:4
Human							
plasma	12	2	27	45		8	
liver	23	10	28	22		6	
Sheep							
plasma	10	2	27	35	7	-	-
liver	17	9	29	7	4	3	-
adrenals	13	7	35	18	2	4	2

In plasma and in the high-density lipoproteins (HDL) in particular, cholesterol esters are synthesised largely by transfer of fatty acids to cholesterol from position sn-2 of phosphatidylcholine ('lecithin') catalysed by the enzyme lecithin:cholesterol acyl transferase (LCAT); the other product is 1-acyl lysophosphatidylcholine. In fact, the reaction occurs in several steps. First, apoprotein A1 in the HDL acts to concentrate the lipid substrates near LCAT and present it in the optimal conformation; at the same time, it opens a lid on the enzyme that activates it by opening up the site of transesterification. Then, cleavage of the sn-2 ester bond of phosphatidylcholine occurs via the phospholipase activity of LCAT with release of a fatty acyl moiety. This is transacylated to the sulfur atom of a cysteine residue forming a thioester, and ultimately it is donated to the 3β-hydroxyl group of cholesterol to form the cholesterol ester. Some LCAT activity has also been detected in apolipoprotein B100-containing particles (β-LCAT activity as opposed to α-LCAT with HDL).

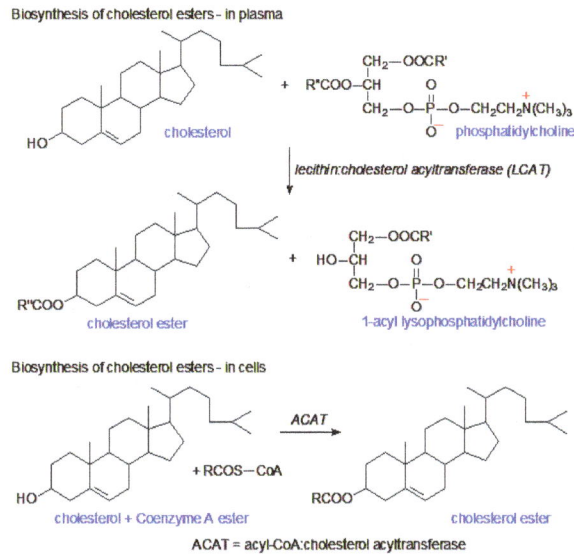

Human LCAT is a relatively small glycoprotein with a polypeptide mass of 49 kDa, increased to about 60 kDa by four *N*-glycosylation and two *O*-glycosylation moieties. Most of the enzyme is produced in the liver and circulates in the blood stream bound reversibly to HDL, where it is activated by the main protein component of HDL, apolipoprotein A1. As cholesterol esters accumulate in the lipoprotein core, cholesterol is removed from its surface thus promoting the flow of cholesterol from cell membranes into HDL. This in turn leads to morphological changes in HDL, which grow and become spherical. Subsequently, cholesterol esters are transferred to the other lipoprotein fractions LDL and VLDL, a reaction catalysed by cholesterol ester transfer protein. This process promotes the efflux of cholesterol from peripheral tissues ('reverse cholesterol transport'), especially from macrophages in the arterial wall, for subsequent delivery to the liver. LCAT is often stated to be the main driving force behind this process, and it is of great importance for cholesterol homeostasis and a suggested target for therapeutic intervention against cardiovascular disease.

The stereospecificity of LCAT changes with molecular species of phosphatidylcholine containing arachidonic or docosahexaenoic acids, when 2-acyl lysophosphatidylcholines are formed. This reaction may be especially important for the supply of these essential fatty acids to the brain in that such lysophospholipids are believed to cross the blood-brain barrier more readily than the free acids.

In other animal tissues, a further enzyme acyl-CoA:cholesterol acyltransferase (ACAT) synthesises cholesterol esters from CoA esters of fatty acids and cholesterol. ACAT exists in two forms, both of which are intracellular enzymes found in the endoplasmic reticulum and possess multiple hydrophobic regions predicted to function as trans-membrane domains; they are members of the membrane-bound *O*-acyltransferase (MBOAT) superfamily. ACAT1 is present in many tissues, but especially in macrophages and adrenal and sebaceous glands, which store cholesterol esters in the form of cytoplasmic lipid droplets; it is responsible for the synthesis of cholesterol esters in arterial foam cells in human atherosclerotic lesions. ACAT2 is found only in the liver and small intestine, and it is believed to be involved in the supply of cholesterol esters to the nascent lipoproteins. Analogous enzymes are found in yeast where ergosterol is the main sterol, but a very different process occurs in plants.

There are suggestions that cholesteryl arachidonate in the lipoprotein LDL is a good substrate for 12/15-lipoxygenase and other oxidizing agents and that the products are a causative factor in atherosclerosis.

Hydrolysis of cholesterol esters: Cholesterol ester hydrolases in animals liberate cholesterol and free fatty acids when required for membrane and lipoprotein formation, and they also provide cholesterol for hormone synthesis in adrenal cells. Many cholesterol ester hydrolases have been identified, including a carboxyl ester hydrolase, a lysosomal acid cholesterol ester lipase, hormone-sensitive lipase and hepatic cytosolic cholesterol ester hydrolase. These are located in many different tissues and organelles and have multiple functions. A neutral cholesterol ester hydrolase has received special study, as it involved in the removal of cholesterol esters from macrophages, so reducing the formation of foam cells and thence the development of fatty streaks within the arterial wall, a key event in the progression of atherosclerosis.

Other Animal Sterols

Cholesterol will oxidize slowly in tissues or foods to form a range of different products with additional hydroperoxy, epoxy, hydroxy or keto groups, and these can enter tissues via the diet. There is increasing interest in these from the standpoint of human health and nutrition, since accumulation of oxo-sterols in plasma is associated with inhibition of the biosynthesis of cholesterol and bile acids and with other abnormalities in plasma lipid metabolism.

A number of other sterols occur in small amounts in tissues, most of which are intermediates in the pathway from lanosterol to cholesterol, although some of them have distinct functions in their own right. Lanosterol, the first sterol intermediate in the biosynthesis of cholesterol, was first found in wool wax, both in free and esterified form, and this is still the main commercial source. It is found at low levels only in most other animal tissues (typically 0.1% of the cholesterol concentration). As oxygen is required, lanosterol cannot be produced by primitive organisms, hence its absence from prokaryotes, leading to some speculation on its evolutionary significance. When sterols became available to eukaryotes, much greater possibilities opened for their continuing evolution. The production of cholesterol from lanosterol is then seen as 'molecular streamlining' by evolution, removing protruding methyl groups that hinder the interaction between sterols and phospholipids in membranes.

Desmosterol (5,24-cholestadien-3β-ol), the last intermediate in the biosynthesis of cholesterol by the Bloch pathway, may be involved in the process of myelination, as it is found in relative abundance in the brains of young animals but not in those of adults, other than astrocytes. It is also found in appreciable amounts in testes and spermatozoa together with another cholesterol intermediate, testis meiosis-activating sterol. In addition, there is evidence that desmosterol activates certain genes involved in lipid biosynthesis in macrophages, and may deactivate others associated with the inflammatory response. There is a rare genetic disorder in which there is an impairment in the conversion of desmosterol to cholesterol, desmosterolosis, with serious consequences in terms of mental capacity. These and related sterols appear to be essential for human reproduction.

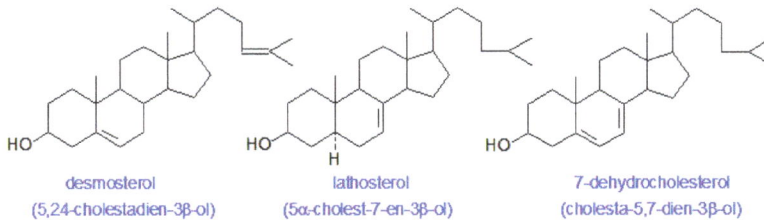

| desmosterol | lathosterol | 7-dehydrocholesterol |
| (5,24-cholestadien-3β-ol) | (5α-cholest-7-en-3β-ol) | (cholesta-5,7-dien-3β-ol) |

In human serum, the levels of lathosterol (5α-cholest-7-en-3β-ol) were found to be inversely related to the size of the bile acid pool, and in general the concentration of serum lathosterol is strongly correlated with the cholesterol balance under most dietary conditions. The isomeric saturated sterols, cholestanol and coprastanol, which differ in the stereochemistry of the hydrogen atom on carbon 5, are formed by microbial biohydrogenation of cholesterol in the intestines, and together with cholesterol are the main sterols in faeces. Further examples of animal sterols include 7-dehydrocholesterol (cholesta-5,7-dien-3β-ol) in the skin, which on irradiation with UV light is converted to vitamin D_3 (cholecalciferol).

Marine invertebrates produce a large number of novel sterols, with both unusual nuclei and unconventional side-chains, some derived from cholesterol and others from plant sterols or alternative biosynthetic intermediates. For example, at least 80 distinct sterols have been isolated from echinoderms and 100 from sponges.

Triacylglycerol Metabolism

All eukaryotic organisms and even a few prokaryotes are able to synthesise triacylglycerols, and in animals, many cell types and organs have this ability, but the liver, intestines and adipose tissue are most active with most of the body stores in the last of these. Within all cell types, even those of the brain, triacylglycerols are stored as cytoplasmic 'lipid droplets' enclosed by a monolayer of phospholipids and hydrophobic proteins such as the perilipins in adipose tissue or oleosins in seeds. These lipid droplets are now treated as distinctive organelles, with their own characteristic metabolic pathways and associated enzymes - no longer boring blobs of fat. However, they are not unique to animals and plants as Mycobacteria and yeasts have similar lipid inclusions.

The lipid serves as a store of fatty acids for energy, which can be released rapidly on demand, and as a reserve of fatty acids for structural purposes or as precursors for eicosanoids. However, lipid droplets may also serve as a protective agency to remove any excess of biologically active and potentially harmful lipids such as free fatty acids, oxylipins, diacylglycerols, cholesterol (as cholesterol esters), retinol esters and coenzyme A esters from cells.

Biosynthesis of Triacylglycerols

Three main pathways for triacylglycerol biosynthesis are known, the *sn*-glycerol-3-phosphate and dihydroxyacetone phosphate pathways, which predominates in liver and adipose tissue, and a monoacylglycerol pathway in the intestines. In maturing plant seeds and some animal tissues, a fourth pathway has been recognized in which a diacylglycerol transferase is involved. The most important route to triacylglycerols is the *sn*-glycerol-3-phosphate or Kennedy pathway, by means of which more than 90% of liver triacylglycerols are produced.

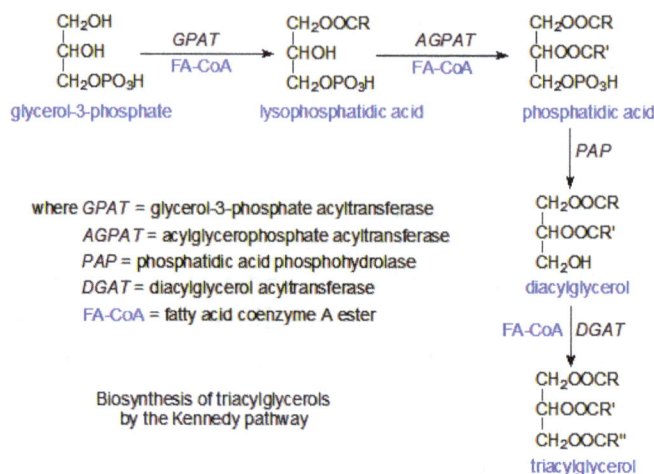

$$
\begin{array}{ccc}
CH_2OH & & CH_2OOCR & & CH_2OOCR \\
| & \xrightarrow[\text{FA-CoA}]{GPAT} & | & \xrightarrow[\text{FA-CoA}]{AGPAT} & | \\
CHOH & & CHOH & & CHOOCR' \\
| & & | & & | \\
CH_2OPO_3H & & CH_2OPO_3H & & CH_2OPO_3H \\
\text{glycerol-3-phosphate} & & \text{lysophosphatidic acid} & & \text{phosphatidic acid}
\end{array}
$$

where *GPAT* = glycerol-3-phosphate acyltransferase
AGPAT = acylglycerophosphate acyltransferase
PAP = phosphatidic acid phosphohydrolase
DGAT = diacylglycerol acyltransferase
FA-CoA = fatty acid coenzyme A ester

$$
\downarrow PAP
$$

$$
\begin{array}{c}
CH_2OOCR \\
| \\
CHOOCR' \\
| \\
CH_2OH \\
\text{diacylglycerol}
\end{array}
$$

$$
\text{FA-CoA} \downarrow DGAT
$$

$$
\begin{array}{c}
CH_2OOCR \\
| \\
CHOOCR' \\
| \\
CH_2OOCR'' \\
\text{triacylglycerol}
\end{array}
$$

Biosynthesis of triacylglycerols
by the Kennedy pathway

In this pathway, the main source of the glycerol backbone has long been believed to be *sn*-glycerol-3-phosphate produced by the catabolism of glucose (glycolysis) or to a lesser extent by the action of the enzyme glycerol kinase on free glycerol. However, there is increasing evidence that a significant proportion of the glycerol is produced *de novo* by a process known as glyceroneogenesis via pyruvate. Indeed, this may be the main source in adipose tissue.

Subsequent reactions occur primarily in the endoplasmic reticulum. First, the precursor *sn*-glycerol-3-phosphate is esterified by a fatty acid coenzyme A ester in a reaction catalysed by a glycerol-3-phosphate acyltransferase (GPAT) at position *sn*-1 to form lysophosphatidic acid, and this is in turn acylated by an acylglycerophosphate acyltransferase (AGPAT) in position *sn*-2 to form a key intermediate in the biosynthesis of all glycerolipids - phosphatidic acid. Numerous isoforms of these enzymes are known; they are expressed with specific tissue and membrane distributions and they are regulated in different ways.

The phosphate group is removed by an enzyme (or family of enzymes) phosphatidic acid phosphohydrolase (PAP or 'phosphatidate phosphatase' or 'lipid phosphate phosphatase'). PAP is also important as it produces *sn*-1,2-diacylglycerols as essential intermediates in the biosynthesis not only of triacylglycerols but also of phosphatidylcholine and phosphatidylethanolamine (and of monogalactosyldiacylglycerols in plants). Much of the phosphatase activity leading to triacylglycerol biosynthesis resides in three related cytoplasmic proteins, termed lipin-1, lipin-2 and lipin-3, which unusually were characterized before the nature of their enzymatic activities were determined. The lipins are tissue specific, and each appears to have distinctive expression and functions, but lipin-1 (PAP1) in three isoforms (designated 1α, 1β and 1γ) accounts for most of the PAP activity in adipose tissue and skeletal muscle. Lipin 2 is the most abundant lipin in liver, but is also expressed substantially in the small intestine, macrophages and some regions of the brain, while lipin 3 activity overlaps with that that of lipin 1 and lipin 2. While it occurs mainly in the cytosolic compartment of cells, it is translocated to the endoplasmic reticulum in response to elevated levels of fatty acids within cells although it does not have trans-membrane domains. Lipin-1 activity requires Mg^{2+} ions and is inhibited by N-ethylmaleimide, whereas the membrane-bound activity responsible for synthesising di-acylglycerols as a phospholipid intermediate is independent of Mg^{2+} concentration and is not sensitive to the inhibitor.

Perhaps surprisingly, lipin-1 has a dual role in that it operates in collaboration with known nuclear receptors as a transcriptional coactivator to modulate lipid metabolism (lipin 1α) while lipin 1β is associated with induction of lipogenic genes such as fatty acid synthase, stearoyl CoA desaturase and DGAT. Abnormalities in lipin-1 expression are known to be involved in some human disease states that may lead to the metabolic syndrome and inflammatory disorders. Lipin 2 is a similar phosphatidate phosphohydrolase, which is present in liver and brain and is regulated dynamically by fasting and obesity (in mice), while lipin 3 is found in the gastrointestinal tract and liver.

In the final step in this pathway, the resultant 1,2-diacyl-sn-glycerol is acylated by diacylglycerol acyltransferases (DGAT), which can utilize a wide range of fatty acyl-CoA esters to form the triacyl-sn-glycerol. In fact there are two DGAT enzymes, which are structurally and functionally distinct. In animals, DGAT1 is located mainly in the endoplasmic reticulum and is expressed in skeletal muscle, skin and intestine, with lower levels of expression in liver and adipose tissue. It is believed to have dual topology contributing to triacylglycerol synthesis on both sides of the membrane of the endoplasmic reticulum, but esterifying only pre-formed fatty acids of exogenous origin DGAT1 is the only one present in the epithelial cells that synthesise milk fat in the mammary gland. Orthologues of this enzyme are present in most eukaryotes, other than yeasts, and it is especially important in plants. Also DGAT1 can utilize a wider range of substrates, including monoacylglycerols, long-chain alcohols (for wax synthesis) and retinol. DGAT2 is the main form of the enzyme in hepatocytes and adipocytes (lipid droplets), although it is expressed much more widely in tissues. It is associated with distinct regions of the endoplasmic reticulum, at the surface of lipid droplets and in mitochondria, and it esterifies fatty acids of both endogenous and exogenous origin. DGAT2 is believed to have a targeting domain that enables it to tether between the endoplasmic reticulum and lipid droplet thereby channelling triacylglycerols from the synthesis site in the endoplasmic reticulum to the nascent lipid droplet, where they accumulate and lead to the expansion of the latter. Both enzymes are important modulators of energy metabolism, although DGAT2 appears to be especially important in controlling the homeostasis of triacylglycerols *in vivo*. As the glycerol-3-phosphate acyltransferase (GPAT) has the lowest specific activity of these enzymes, this step may be the rate-limiting one. However, DGATs are the dedicated triacylglycerol-forming enzymes, and they are seen as the best target for pharmaceutical intervention in obesity and attendant ailments; clinical studies of DGAT1 inhibitors are at an early stage.

In a second pathway for triacylglycerol biosynthesis, dihydroxyacetone-phosphate in peroxisomes or endoplasmic reticulum can be acylated by a specific acyltransferase to form 1-acyl dihydroxyacetone-phosphate, which is reduced by dihydroxyacetone-phosphate oxido-reductase to lysophosphatidic acid, which can then enter the pathway above to triacylglycerols. The precursor dihydroxyacetone-phosphate is important also as part of the biosynthetic route to plasmalogens, and neutral plasmalogens can be significant components of cytoplasmic droplets in many mammalian cells types but not adipose tissue.

$$
\begin{array}{c}
\text{CH}_2\text{OH} \\
| \\
\text{C}=\text{O} \\
| \\
\text{CH}_2\text{OPO}_3\text{H}
\end{array}
\xrightarrow{\text{FA-CoA}}
\begin{array}{c}
\text{CH}_2\text{OOCR} \\
| \\
\text{C}=\text{OH} \\
| \\
\text{CH}_2\text{OPO}_3\text{H}
\end{array}
\longrightarrow
\begin{array}{c}
\text{CH}_2\text{OOCR} \\
| \\
\text{CHOH} \\
| \\
\text{CH}_2\text{OPO}_3\text{H}
\end{array}
\longrightarrow
\text{phosphatidic acid}
$$

dihydroxyacetone-phosphate   1-acyl dihydroxyacetone-phosphate   lysophosphatidic acid   sn-1,2-diacylglycerols

↓

triacylglycerols

Biosynthesis of triacylglycerols via dihydroxyacetone-phosphate

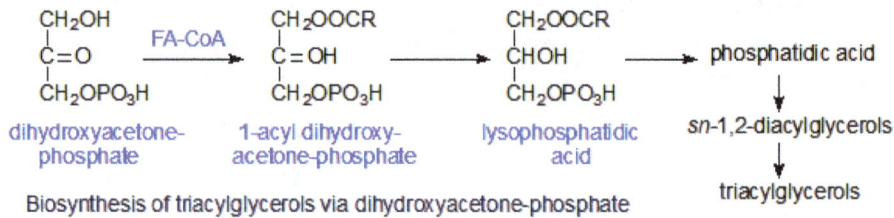

In the enterocytes of intestines after a meal, up to 75% of the triacylglycerols are formed via a monoacylglycerol pathway. 2-Monoacyl-sn-glycerols and free fatty acids released from dietary triacylglycerols by the action of pancreatic lipase within the intestines (see below) are taken up by the enterocytes. There, the monoacylglycerols are first acylated by an acyl coenzyme A:monoacylglycerol acyltransferase with formation of sn-1,2-diacylglycerols mainly as the first intermediate in the process, though some sn-2,3-diacylglycerols (~10%) are also produced (DGAT1 can also acylate monoacylglycerols). 1-Monoacylglycerols can also be synthesised by the acylation of glycerol and these can also be acylated. There are three isoforms of the monoacylglycerol acyltransferase in humans of which MGAT2 is most active in the intestines and liver and MGAT1 in adipose tissue. Finally, the acyl coenzyme A:diacylglycerol acyltransferase (DGAT1) reacts with the sn-1,2-diacylglycerols only to form triacylglycerols.

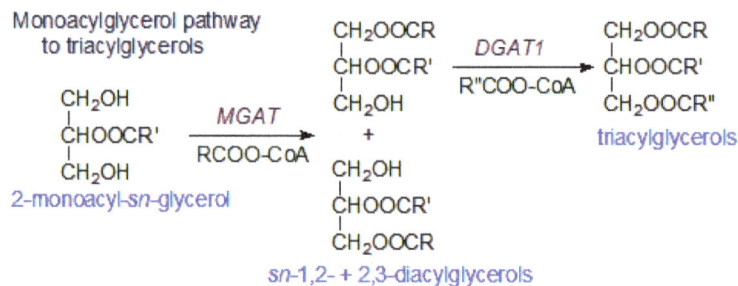

Monoacylglycerol pathway to triacylglycerols

$$
\begin{array}{c}
\text{CH}_2\text{OH} \\
| \\
\text{CHOOCR'} \\
| \\
\text{CH}_2\text{OH}
\end{array}
\xrightarrow[\text{RCOO-CoA}]{\text{MGAT}}
\begin{array}{c}
\text{CH}_2\text{OOCR} \\
| \\
\text{CHOOCR'} \\
| \\
\text{CH}_2\text{OH} \\
+ \\
\text{CH}_2\text{OH} \\
| \\
\text{CHOOCR'} \\
| \\
\text{CH}_2\text{OOCR}
\end{array}
\xrightarrow[\text{R"COO-CoA}]{\text{DGAT1}}
\begin{array}{c}
\text{CH}_2\text{OOCR} \\
| \\
\text{CHOOCR'} \\
| \\
\text{CH}_2\text{OOCR"}
\end{array}
$$

2-monoacyl-sn-glycerol   sn-1,2- + 2,3-diacylglycerols   triacylglycerols

In a fourth biosynthetic pathway, which is less well known, triacylglycerols are synthesised by a transacylation reaction between two racemic diacylglycerols that is independent of acyl-CoA. The reaction was first detected in the endoplasmic reticulum of intestinal micro villus cells and is catalysed by a diacylglycerol transacylase. Both diacylglycerol enantiomers participate in the reaction with equal facility to transfer a fatty acyl group with formation of triacylglycerols and a 2-monoacyl-sn-glycerol. A similar reaction has been observed in seed oils.

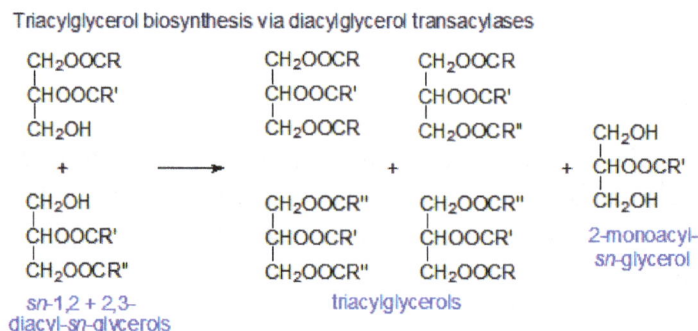

Triacylglycerol biosynthesis via diacylglycerol transacylases

$$
\begin{array}{c}
\text{CH}_2\text{OOCR} \\
| \\
\text{CHOOCR'} \\
| \\
\text{CH}_2\text{OH} \\
+ \\
\text{CH}_2\text{OH} \\
| \\
\text{CHOOCR'} \\
| \\
\text{CH}_2\text{OOCR"}
\end{array}
\longrightarrow
\begin{array}{c}
\text{CH}_2\text{OOCR} \\
| \\
\text{CHOOCR'} \\
| \\
\text{CH}_2\text{OOCR"} \\
+ \\
\text{CH}_2\text{OOCR"} \\
| \\
\text{CHOOCR'} \\
| \\
\text{CH}_2\text{OOCR"}
\end{array}
\quad
\begin{array}{c}
\text{CH}_2\text{OOCR} \\
| \\
\text{CHOOCR'} \\
| \\
\text{CH}_2\text{OOCR"} \\
+ \\
\text{CH}_2\text{OOCR"} \\
| \\
\text{CHOOCR'} \\
| \\
\text{CH}_2\text{OOCR}
\end{array}
\quad
+
\begin{array}{c}
\text{CH}_2\text{OH} \\
| \\
\text{CHOOCR'} \\
| \\
\text{CH}_2\text{OH}
\end{array}
$$

sn-1,2 + 2,3-diacyl-sn-glycerols   triacylglycerols   2-monoacyl-sn-glycerol

This enzyme may function in remodelling triacylglycerols post synthesis, especially in oil seeds, and it is possible that it may be involved in similar processes in the liver and adipose tissue, where extensive hydrolysis/re-esterification is known to occur. There is evidence for selectivity in the biosynthesis of different molecular species in a variety of tissues and organisms, which may be a consequence of the varying biosynthetic pathways. Also in adipose tissue, fatty acids synthesised *de novo* are utilized in different ways from those from external sources in that they enter positions *sn*-1 and 2 predominantly, while a high proportion of the oleic acid synthesised in the tissue by desaturation of exogenous stearic acid is esterified to position *sn*-3.

In prokaryotes, the glycerol-3-phosphate pathway of triacylglycerol biosynthesis only occurs, but in yeast both glycerol-3-phosphate and dihydroxyacetone-phosphate can be the primary precursors and synthesis takes place in cytoplasmic lipid droplets and the endoplasmic reticulum. In plants, the glycerol-3-phosphate pathway is most important.

Among other potential routes to the various intermediates, lysophosphatidic acid and phosphatidic acid can be synthesised in mitochondria, but must then be transported to the endoplasmic reticulum before they enter the pathway for triacylglycerol production. 1,2-Diacyl-*sn*-glycerols are also produced by the action of phospholipase C on phospholipids.

In the glycerol-3-phosphate and other pathways, the starting material is of defined stereochemistry and each of the enzymes catalysing the various steps in the process is distinctive and can have preferences for particular fatty acids (as their coenzyme A esters) and for particular fatty acid combinations in the partially acylated intermediates. It should not be surprising, therefore, that natural triacylglycerols exist in enantiomeric forms with each position of the *sn*-glycerol moiety esterified by different fatty acids.

While triacylglycerols are essential for normal physiology, an excessive accumulation in human adipose tissue and other organs results in obesity and other health problems, including insulin resistance, steatohepatitis and cardiomyopathy. Accordingly, there is considerable pharmaceutical interest in drugs that affect triacylglycerol biosynthesis and metabolism.

Triacylglycerol Metabolism in Humans

The process of fat digestion is begun in the stomach by acid-stable gastric or lingual lipases, the extent of which depending on species but may be important for efficient emulsification. However, this is insignificant in quantitative terms in comparison to the reaction with pancreatic lipase, which occurs in the duodenum. Entry of triacylglycerol degradation products into the duodenum stimulates synthesis of the hormone cholecystokinin and causes the gall bladder to release bile acids, which are strong detergents and act to emulsify the hydrophobic triacylglycerols so increasing the available surface area. In turn, cholecystokinin stimulates the release of the hydrolytic enzyme pancreatic lipase together with a co-lipase, which is essential for the activity of the enzyme. Pancreatic lipase, co-lipase, bile salts and calcium ions act together in a complex at the surface of the emulsified fat droplets to hydrolyse triacylglycerols. The process is regiospecific and results in the release of the fatty acids from the 1 and 3 positions with formation of 2-monoacyl-*sn*-glycerols. Isomerization of the latter to 1(3)-monoacyl-*sn*-glycerols occurs to some extent, and these can be degraded completely by the enzyme to glycerol and free fatty acids. Other lipases hydrolyse the phospholipids and other complex lipids in foods at the same time.

Hydrolysis of triacylglycerols in the intestines

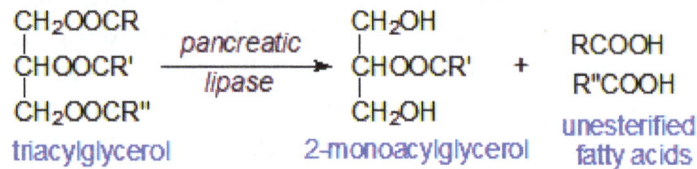

$$
\begin{array}{c}
\text{CH}_2\text{OOCR} \\
| \\
\text{CHOOCR'} \\
| \\
\text{CH}_2\text{OOCR''} \\
\text{triacylglycerol}
\end{array}
\quad
\xrightarrow[\text{lipase}]{\textit{pancreatic}}
\quad
\begin{array}{c}
\text{CH}_2\text{OH} \\
| \\
\text{CHOOCR'} \\
| \\
\text{CH}_2\text{OH} \\
\text{2-monoacylglycerol}
\end{array}
\quad + \quad
\begin{array}{c}
\text{RCOOH} \\
\text{R''COOH} \\
\text{unesterified} \\
\text{fatty acids}
\end{array}
$$

This process is somewhat different in neonates and young infants, in whom pancreatic lipase is less active but is effectively replaced by lipases in breast milk and by an acid gastric lipase (pH optimum 4-6).

There is evidence that the regiospecific structure of dietary triacylglycerols has an effect on the uptake of particular fatty acids and may influence further the lipid metabolism in humans. In particular, incorporation of palmitic acid into the position sn-2 of milk fat may be of benefit to the human infant (as a source of energy for growth and development), although it increases the atherogenic potential for adults. In addition, 2-monoacylglycerols and 2-oleoylglycerol especially have a signalling function in the intestines by activating a specific G-protein coupled receptor GPR119, sometimes termed the 'fat sensor'. When stimulated, this causes a reduction in food intake and body weight gain in rats and regulates glucose-stimulated insulin secretion. The free fatty acids released have a similar effect, though by a very different mechanism, via the receptor GPR40. Overall, it has become evident that triacylglycerol metabolism in the intestine has regulatory effects on the secretion of gut hormones and on systemic lipid metabolism and energy balance.

The free fatty acids and 2-monoacyl-sn-glycerols are rapidly taken up by the intestinal cells, from the distal duodenum to the jejunum, via specific carrier molecules but also by passive diffusion. A specific fatty acid binding protein prevents a potentially toxic build-up of unesterified fatty acids and targets them for triacylglycerol biosynthesis. The long-chain fatty acids are converted to the CoA esters and esterified into triacylglycerols by the monoacylglycerol pathway as described above. In contrast, short and medium-chain fatty acids (C_{12} and below) are absorbed in unesterified form and pass directly into the portal blood stream, where they are transported to the liver to be oxidized.

Subsequently, the triacylglycerols are incorporated into lipoprotein complexes termed chylomicrons in the enterocytes by processes. In brief, these consist of a core of triacylglycerols together with some cholesterol esters that is stabilized and rendered compatible with an aqueous environment by a surface film consisting of phospholipids, free cholesterol and one molecule of a truncated form of apoprotein B (48 kDa). These particles are secreted into the lymph and thence into the plasma for transport to the peripheral tissues for storage or structural purposes. Adipose tissue in particular exports appreciable amounts of the enzyme lipoprotein lipase, which binds to the luminal membrane of endothelial cells facing into the blood, where it rapidly hydrolyses the passing triacylglycerols at the cell surface releasing free fatty acids, most of which are absorbed into the adjacent adipocytes and re-utilized for triacylglycerol synthesis within the cell.

The chylomicrons remnants eventually reach the liver, where the remaining lipids are hydrolysed at the external membranes by a hepatic lipase and absorbed. The fatty acids within the liver can be utilized for a variety of purposes, from oxidation to the synthesis of structural lipids, but a proportion is re-converted into triacylglycerols, and some of this is stored as lipid droplets within the cytoplasm of the cells. In addition, phosphatidylcholine from the high-density lipoproteins is taken up by the liver, and a high proportion of this is eventually converted to triacylglycerols. In healthy liver, the levels of triacylglycerols are low (<5% of the total lipids), because the rates of acquisition of fatty acid from plasma and synthesis *de novo* within the liver are balanced by rates of oxidation and secretion into plasma. On the other hand, excessive accumulation of storage triacylglycerols is associated with fatty liver, insulin resistance and type 2 diabetes.

Most of the newly synthesised triacylglycerols are exported into the plasma in the form of very-low-density lipoproteins (VLDL), consisting again of a triacylglycerol and cholesterol ester core, surrounded by phospholipids and free cholesterol, together with one molecule of full-length apoprotein B (100 kDa), apoprotein C and sometimes apoprotein E. These particles in turn are transported to the peripheral tissues, where they are hydrolysed and the free acids absorbed. Eventually, the remnants are returned to the liver.

In the mammary gland, triacylglycerols are synthesised in the endoplasmic reticulum and large lipid droplets are produced with a monolayer of phospholipids derived from this membrane. These are transported to the plasma membrane and bud off into the milk with an envelope comprised of the phospholipid membrane to form milk fat globules as food for the newborn. The process is thus very different from that involved in the secretion of triacylglycerol-rich lipoproteins from other organs.

Triacylglycerol Synthesis and Catabolism (Lipolysis) in Adipocytes and Lipid Droplets

Adipose tissue and the adipocytes are characterized by accumulations of triacylglycerols, which act as the main energy store for animals, although they also cushion and insulate the body. Thus, triacylglycerols stored when there is a surplus of nutrients are mobilized for energy production during starvation. Adipose tissue also functions as a reserve of bioactive lipids, such as eicosanoids and lipid-soluble vitamins, and when required provides structural components, including fatty acids, cholesterol and retinol, for membrane synthesis and repair. Large depots occur around internal organs such as the liver, and also subcutaneously. Brown and beige fat have special properties and are, while bone marrow adipocytes (70% of the available space) have distinctive functions also.

Similarly, within most other animal cells, even ganglia in the brain, a proportion of the fatty acids taken up from the circulation is converted to triacylglycerols as described above and incorporated into cytoplasmic lipid droplets (also termed 'fat globules', 'oil bodies', 'lipid particles', 'adiposomes', etc). By buffering against fatty acid accumulation that might exceed the capacity of non-adipose cells, they defend them against lipotoxicity while providing a rapid source of energy and essential metabolites. Acting in concert with other cellular organelles, they function in many different metabolic processes. The triacylglycerol droplets are surrounded by a protective monolayer that includes phospholipids, cholesterol and hydrophobic proteins. The phospholipid component of the monolayer consists mainly of phosphatidylcholine and phosphatidylethanolamine with fatty acid compositions distinct from those of the endoplasmic reticulum and plasma membrane.

Among the proteins are many that function directly in lipid metabolism, and they include acyltransferases, lipases, perilipins, caveolins and the Adipose Differentiation Related Protein (ADRP or adipophilin). In adipocytes, the lipid droplets can range to up to 200 μm in diameter, while other cell types contain smaller lipid droplets of the order of 50 nm in diameter. Cytosolic lipid droplets with similar metabolic activities are found in the fly *Drosophila melanogaster*, and in higher plants and yeasts. Like adipose tissue cells, lipid droplets have a major function in that they sense and respond rapidly to changes in systemic energy balance. Within cells, lipid droplets facilitate the coordination and communication between different organelles and act as vital hubs of cellular metabolism. They secrete important hormone-like molecules such as leptin, adiponectin and adipsin, and so influence food intake, insulin sensitivity, insulin secretion and related processes.

Lipid droplet assembly - This process takes place in the endoplasmic reticulum, where at least one isoform of each of the enzymes of triacylglycerol biosynthesis, from acyl-CoA synthetases through to glycerol-3-phosphate acyltransferases, is located probably in a protein assembly or 'interactome'. Triacylglycerols accumulate and so attract perilipins and other proteins that allow lipid droplets to grow as a lens-like swelling in patches of the membrane. A protein seipin stabilizes the nascent droplets with minimal disruption to the membrane and enables them to mature by a mechanism that is still uncertain but may involve regulation of protein and lipid trafficking into the droplet. The growing lipid droplets bud toward the cytosol, a process that is believed to be directed and aided by surface proteins such as perilipin, while the triacylglycerol core attracts and is largely surrounded by phospholipids from the outer leaflet of the endoplasmic reticulum. Growth continues through an extended endoplasmic reticulum/droplet junction or bridge until finally with the involvement of membrane curvatureinducing coat proteins, trans-membrane ('FIT') proteins that bind diacylglycerol intermediates and specific phospholipid species, the droplets bud off into the cytoplasm with their surface monolayer of phospholipids and proteins, including the enzymes of triacylglycerol biosynthesis. Subsequently, mitochondria, peroxisomes and other organelles may contribute or exchange lipids and effect changes in protein composition, although the lipid droplets remain close to the endoplasmic reticulum, and presumably this enables a dynamic response to any change in metabolic status sensed by the this organelle. In adipocytes, lipid droplets can also grow by fusion of smaller droplets, although again the mechanism is not fully understood.

Some of the surface proteins on lipid droplets can extend long helical hairpins of hydrophobic peptides deep into the lipid core. For example, perilipins constitute a family of at least five phosphorylated proteins that bind to droplets in animals and share a common region, the so-called 'PAT' domain, named for the three original members of the family that include perilipin and ADRP. Proteins related evolutionarily to these are found in more primitive organisms, including insects, slime moulds and fungi, but not in the nematode *Caenorhabditis elegans*. In mammals, perilipin A (or 'PLIN1' or more accurately the splice variant 'PLIN1a') is a well-established regulator of lipolysis in adipocytes, and it is believed to be involved in the formation of the large lipid droplets in white adipose tissue. The perilipins PLIN1 and PLIN2 have functions in triacylglycerol metabolism in tissues other than adipocytes, and PLIN2 in particular is the main perilipin in hepatocytes; PLIN5 operates in tissues that oxidize fatty acids such as the heart. Other surface proteins of lipid droplets are enzymes intimately involved in triacylglycerol metabolism, although there is a suggestion that cytoplasmic droplets may act as a storage organelle for hydrophobic proteins whose function is elsewhere in the cell.

On the basis of profiling of the surface proteins and phospholipids, lipid droplets in cells are now considered to be complex, metabolically active organelles that function in the supply of fatty acids for various purposes, including membrane trafficking and possibly in the recycling of both simple and complex lipids. Although they originate in the endoplasmic reticulum, lipid droplets can associate with most other cellular organelles through membrane contact sites in a highly dynamic manner. For example, within the liver, triacylglycerols are stored as lipid droplets in the cytoplasm adjacent to the endoplasmic reticulum where a triacylglycerol hydrolase can effect lipolysis to di- and monoacylglycerols that are more soluble in the membrane, which they are able to cross. They are then available for re-synthesis into triacylglycerols by luminally oriented acyltransferases before assembly into nascent lipoprotein complexes. Similar organelles can be found in most eukaryotic cells and in bacteria, and they provide a reservoir not only for triacylglycerols and their fatty acid constituents but also for eicosanoids, and cholesterol and retinol esters, for example. Lipid droplets even occur in the nuclei of cells where their constituents interact with numerous proteins and are involved in the regulation of nuclear events.

Lipolysis - When fatty acids are required by other tissues for energy or other purposes, they are released from the triacylglycerols by the sequential actions of three enzymes, i.e. adipose triacylglycerol lipase (ATGL), hormone-sensitive lipase (HSL) and monoacylglycerol lipase, which cycle between the cytoplasmic surfaces of the endoplasmic reticulum and the surface layer of lipid droplets. Simplistically, ATGL hydrolyses triacylglycerols to diacylglycerols, which are hydrolysed by HSL to monoacylglycerols before these are hydrolysed by the monoacylglycerol lipase. Perilipin (PLIN1) has been described as "the gatekeeper of the adipocyte lipid storehouse". Thus, the lipolytic process is regulated by perilipin, which acts as a barrier to lipolysis in non-stimulated cells, but on stimulation as during fasting is phosphorylated by the cAMP-protein kinase. This changes its shape and reduces its hydrophobicity, and in the process activates lipolysis. An isoform, perilipin A, is the main regulatory factor in white adipose tissue.

The adipose triacylglycerol lipase, which initiates the process, was discovered surprisingly recently. It is structurally related to the plant acyl-hydrolases in that it has a patatin-like domain in the NH_2-terminal region (patatin is a non-specific acyl-hydrolase in potato) and is located on the surface of the lipid droplet both in the basal and activated states. This lipase is specific for triacylglycerols containing long-chain fatty acids, preferentially cleaving ester bonds in the sn-1 or sn-2 position (but not sn-3), and yields diacylglycerols and free fatty acids as the main products, with low activity only towards diacylglycerols, and none to monoacylglycerols and cholesterol esters. However, it also has transacylase and phospholipase activities, and it hydrolyses retinol esters in hepatic stellate cells. Adipose triacylglycerol lipase can be activated at the same time as hormone-sensitive lipase and is now believed to be rate limiting for the first step in triacylglycerol hydrolysis. Regulation of the enzymatic activity is a complex process, and for example, a lipid droplet protein designated 'CGI-58' or 'ABHD5', is known to be an important activating factor and is required for hydrolysis of fatty acids from position sn-1. In the resting state this protein binds to perilipin, but on phosphorylation of the latter, it dissociates and interacts with adipose triacylglycerol lipase to activate triacylglycerol hydrolysis. Mutations in adipose triacylglycerol lipase or CGI-58 are believed to be responsible for a syndrome in humans known as 'neutral lipid storage disease'. In contrast, a second protein (G0S2) inhibits the enzyme.

Hormone-sensitive lipase is regulated by the action of the hormones insulin and noradrenaline by a mechanism that ultimately involves phosphorylation of the enzyme by cAMP-protein kinase (as with perilipin), thereby increasing its activity and causing it to translocate from the cytosol to the lipid droplet. In addition to its activity towards triacylglycerols, hormone-sensitive lipase will rapidly hydrolyse diacylglycerols, monoacylglycerols, retinol esters and cholesterol esters. In fact, diacylglycerols are hydrolysed ten times as rapidly as triacylglycerols. Within the triacylglycerol molecule, hormone-sensitive lipase preferentially hydrolyses ester bonds in the *sn*-1 and *sn*-3 positions, leaving free acids and 2-monoacylglycerols as the main end products.

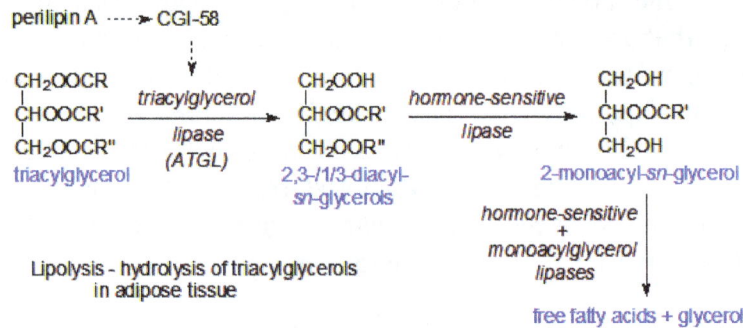

Lipolysis - hydrolysis of triacylglycerols in adipose tissue

The monoacylglycerol lipase is believed to be the rate-limiting enzyme in monoacylglycerol hydrolysis, i.e. the final step in triacylglycerol catabolism releasing free glycerol and fatty acids, and is found in the cytoplasm, the plasma membrane, and in lipid droplets. It is specific for monoacylglycerols and has no activity against di- or triacylglycerols. As it is the enzyme mainly responsible for deactivation of the endocannabinoid 2-arachidonoylglycerol and is highly active in malignant cancers, it is attracting pharmaceutical interest. Further lipolytic enzymes, including carboxylesterases, are believed to operate against triacylglycerols in cytoplasmic lipid droplets in the liver.

Free fatty acids released by the combined action of these enzymes are exported into the plasma for transport to other tissues in the form of albumin complexes, while the glycerol released is transported to the liver for metabolism by either glycolysis or gluconeogenesis. Eventually, the whole organelle can disappear, including the proteins, when they undergo a process of autophagy ('lipophagy'), i.e. the delivery of the organelles to lytic compartments for degradation. This process is especially important during starvation and is also relevant to tumorigenesis and cancer metastasis, and while it is mechanistically distinct from lipolysis, there is cross-talk between the two.

Not only does the adipocyte provide a store of energy but it manages the flow of energy through the formation of the hormone leptin, which stimulates secretion of various factors that communicate with other tissues, including cytokines, adiponectin and resistin. The synthesis of leptin is tightly controlled by adipocytes and its main function is believed to be the provision of information on the state of fat stores to other tissues. Lipid droplets may play a role in this process, since perilipin is required for the sensing function. As caveolae, which contain the proteins caveolins (and presumably sphingolipids) and are particularly abundant in adipocytes, modulate the flux of fatty acids across the plasma membrane and are involved in signal transduction and membrane trafficking pathways, it is evident that they have a major role in this aspect of lipid metabolism. In addition, insulin is the main hormone that affects metabolism and its receptor at the plasma membrane is located in caveolae. Release of proinflammatory cytokines can stimulate lipolysis and cause insulin resistance, in turn leading to dysfunction of adipose tissue and systemic disruption of metabolism. Thus, adipose tissue metabolism has profound effects on whole-body metabolism, and defects in these processes can have severe implications for the pathogenesis of diabetes and obesity in humans.

Functions other than energy management: Lipid droplets accumulate within many cell types other than adipocytes, including leukocytes, epithelial cells and hepatocytes, especially during infectious, neoplastic and other inflammatory conditions. They are important for the cellular storage and release of hydrophobic vitamins, signalling precursors and other lipids that are not related to energy homeostasis. These are associated with a variety of enzymes, including protein kinases, which are involved in many different aspects of lipid metabolism, including cell signalling, membrane trafficking and control of the production of inflammatory mediators such as eicosanoids. For example, the triacylglycerols in cytoplasmic lipid droplets of human mast cells have a high content of arachidonic acid, which can be released by adipose triacylglycerol lipase as a substrate for eicosanoid production. Similarly, triacylglycerols in lipid droplets of the skin are a highly specific source of the linoleic acid that is required for the formation of the O-acylceramides, which are essential for epidermal barrier function. Vitamin E (tocopherols) and vitamin A in the form of retinyl esters are stored in cytoplasmic lipid droplets, and the latter are present in appreciable concentrations in the stellate cells of liver, for example. In endocrine cells of the gonads and adrenals, cholesterol esters stored in lipid droplets are an important source of cholesterol for the mitochondrial biosynthesis of various steroid hormones. In the nucleus, they can sequester transcription factors and chromatin components and generate the lipid ligands for certain nuclear receptors.

In addition to their role in lipid biochemistry, lipid droplets participate in protein degradation and glycosylation. Their metabolism can be manipulated by pathogenic viruses and bacteria such as *Mycobacterium tuberculosis* with unfortunate consequences for the host, but they also serve as reservoirs for proteins that fight intracellular pathogens. In consequence, such lipid droplets and their enzyme systems may be markers for disease states and are also considered to be targets for pharmaceutical intervention.

Subcutaneous depots act as a cushion around joints and serve as insulation against cold in many terrestrial animals, as is obvious in the pig, which is surrounded by a layer of fat, and it is especially true for marine mammals such as seals. Those adipocytes embedded in the skin differ from the general subcutaneous depots and support the growth of hair follicles and regenerating skin, and they may also have a defensive role both as a physical barrier and by responding metabolically to bacterial infection.

In marine mammals and fish, the fat depots are less dense than water and so aid buoyancy with the result that less energy is expended in swimming. More surprisingly perhaps, triacylglycerols together with the structurally related glyceryl ether diesters and wax esters are the main components of the sonar lens used in echo-location by dolphins and toothed whales. The triacylglycerols are distinctive in that they contain two molecules of 3-methylbutyric (isovaleric) acid with one long-chain fatty acid. It appears that the relative concentrations of the various lipids in an organ in the head of the animals (termed the 'melon') are arranged anatomically in a three-dimensional topographical pattern to enable them to focus sound waves.

In cold climates, many insects do not feed over winter and must manage their energy stores to meet the energetic demands of development and reproduction in the spring. Some insect species that are tolerant of freezing produce triacylglycerols containing acetic acid, and these remain liquid at low temperatures; by interacting with water, they may play a role in cryoprotection.

Triacylglycerol Metabolism in Plants and Yeasts

Seed oils - Fruit and seed oils are major agricultural products with appreciable economic and nutritional value to humans. The mesocarp of fruits is a highly nutritious energy source that attracts animals that help to disperse the seeds, and in plants such as the oil palm and olives a high proportion of the flesh contains triacylglycerols. Similarly in seeds, triacylglycerols are the main storage lipid and can comprise as much as 60% of their weight.

In plants, fatty acids synthesised in the plastid compartment are stored in the embryo or endosperm tissues of seeds as triacylglycerols in lipid droplets with a coherent surface layer of proteins and lipids. In addition to the common range of fatty acids synthesised in plastids, some plant species produce novel fatty acids, including medium- and very-long-chain components and those with oxygenated and other functional moieties. Some very specific means of diverting these to seeds and triacylglycerol production must exist to prevent disruption of the plant membranes. Seed development occurs in three stages - rapid cell division with no accumulation of storage material, rapid deposition of triacylglycerols and other energy-rich metabolites, and finally desiccation. In comparison, relatively little triacylglycerol biosynthesis and metabolism occurs in tissues other than developing seeds.

During the period of oil accumulation in seeds, there must be a mechanism to hydrolyse the newly formed ACP esters of fatty acids and export the unesterified fatty acids to the endoplasmic reticulum, where they are converted to the CoA esters and triacylglycerols are synthesised by the Kennedy and other pathways. In yeast and plants, diacylglycerol esterification is the only committed step in triacylglycerol production and this occurs by mechanisms that can be both dependent or independent of acyl-CoA esters. The acyl-CoA-dependent route is catalysed by diacylglycerol:acyl-CoA acyltransferases (DGATs) with acyl-CoA and diacylglycerols as substrates. In plants, two membrane-bound enzymes (DGAT1 and DGAT2) and a cytosolic enzyme (DGAT3) are known, while the acyl-CoA-independent reaction utilizes a phospholipid:diacylglycerol acyltransferase with phospholipids as acyl donor and diacylglycerol as acyl acceptor to produce triacylglycerols and lysophospholipids. DGAT2 is especially important in those plant species with unusual fatty acid compositions.

In addition, a substantial proportion in some species is synthesised by a flux through the membrane phospholipid phosphatidylcholine, produced by both the eukaryotic and prokaryotic pathways with differing positional distributions, in which diacylglycerols are generated by the action of a phosphatidate phosphatase as an intermediate. As phosphatidylcholine undergoes extensive remodelling and its fatty acid components are subject to modification, for example by desaturation to form linoleic and linolenic acids, the compositions and especially the positional distributions of triacylglycerols produced in this way can be very different from those synthesised by the 'classical' pathways. Phosphatidylcholine may also function as a carrier for the trafficking of acyl groups between organelles and membrane subdomains. As triacylglycerol synthesis continues, oil droplets accumulate in the endoplasmic reticulum and are surrounded by a monolayer of phospholipids and proteins, which in Arabidopsis include oleosins, a caleosin, a steroleosin, a putative aquaporin and a glycosylphosphatidylinositol-anchored protein.

At the onset of germination, water is absorbed and lipases are activated. The process of lipolysis begins at the surface of oil bodies, where the oleosins, which are the most abundant structural proteins, are believed to serve to assist the docking of lipases. They also control the size and stability of lipid droplets in seeds. A number of lipases have been cloned from various plant species and are typical α/β-hydrolases, with a conserved catalytic triad of Ser, His, and Asp or Glu as in patatin (an especially abundant lipolytic protein in potatoes), which are able to hydrolyse triacylglycerols but not phospho- or galactolipids. The most important of these is believed to be the 'sugar-dependent lipase 1 (SDP1)', which is a patatin-like lipase similar to the mammalian adipose triacylglycerol lipase discussed above, and is located on the surface of the oil body. This is active mainly against triacylglycerols to generate diacylglycerols, but presumably works in conjunction with di- and monoacylglycerol lipases to generate free fatty acids and glycerol.

The lipid droplets in seeds exist in close proximity with glyoxysomes (broadly equivalent to peroxisomes). These are the membrane-bound organelles that contain most of the enzymes required to oxidize fatty acids derived from the triacylglycerols via acetyl-CoA to four-carbon compounds, such as succinate, which are then converted to soluble sugars to provide germinating seeds with energy to fuel the growth of the seedlings and to produce shoots and leaves. In addition, they supply structural elements before the seedlings develop the capacity to photosynthesise. The free acids are converted to their coenzyme A esters by two long-chain acyl-CoA synthetases located on the inner face of the peroxisome membrane before entry into the β-oxidation pathway. All of these processes are controlled by an intricate regulatory network, involving transcription factors that crosstalk with signalling events from the seed maturation phase through to embryo development. After about two days of the germination process, the glycoxysomes begin to break down, but β-oxidation can continue in peroxisomes in leaf tissue.

Lipid droplets - plastoglobules: Triacylglycerol-rich lipid droplets (LD) have been observed in most cell types in vegetative tissues of plants as well as in seeds, and although their origin and function are poorly understood, they contain all the enzymes required for triacylglycerol metabolism together with phospholipases, lipoxygenases and other oxidative enzymes. Rather than oleosins, these lipid droplets in plants and algae contain a family of ubiquitously expressed 'LD-associated proteins' on the surface, together with a monolayer of phospholipids (mainly phosphatidylcholine), galactolipids such as sulfoquinovosyldiacylglycerol and in some species betaine lipids. LD form within the bilayer of the endoplasmic reticulum and pinch off into the cytoplasm. While they are believed to be involved mainly in stress responses, they may have other specialized roles, for example in anther and pollen development, where triacylglycerols serve as a source of fatty acids for membrane biosynthesis. Fatty acids derived from triacylglycerols in lipid droplets are believed to be subjected to peroxisomal β-oxidation to produce the ATP required for stomatal opening and no doubt many other purposes.

In addition, lipid droplets that have been termed 'plastoglobules' are produced by localized accumulation of triacylglycerols and other neutral lipids between the membrane leaflets of the thylakoid cisternae and then pinch off into the stroma, where they are involved in a wide range of biological functions from biogenesis to senescence via the recruitment of specific proteins. During senescence for example, lipid droplets accumulate rapidly in leaves of *A. thaliana*. In reproductive tissues, may have a more direct function by recruiting and transporting proteins, both for organ formation and successful pollination. Antifungal compounds such as 2-hydroxy-octadecatrienoic acid are produced from α-linolenic acid in these organelles, and it has been suggested that they function as intracellular factories to produce stable metabolites via unstable intermediates by concentrating the enzymes and hydrophobic substrates in an efficient manner. Plastoglobules are also implicated in the biosynthesis and metabolism of vitamins E and K.

Microalgae and yeasts - Triacylglycerol metabolism in lipid droplets in microalgae is under intensive study because of their potential for biodiesel production. It seems that similar processes occur as in higher plants, but with a simpler genome encoding few redundant proteins. The size and triacylglycerol content of lipid droplets in yeasts change appreciably in different stages of growth and development. As most of the important biosynthetic and catabolic enzymes involved in triacylglycerol metabolism are conserved between yeasts (e.g. *Saccharomyces cerevisiae*) and mammals, the former are proving to be useful models for the study of triacylglycerol homeostasis. Indeed, lipid droplets in yeast are considered to be a highly dynamic and functionally diverse hub that ensures stress resistance and cell survival by promoting membrane and organelle homeostasis.

References

- Lipid, science: britannica.com , Retrieved January 10, 2019

- Fatty_acids, undervisningsmateriale, nedlagte-emner, farmasi, matnat, emner: uio.no, Retrieved May 23, 2019

- Fatty-acid-oxidation-and-synthesis, metabolism-and-hormones: diapedia.org, Retrieved August 16, 2019

- Lipoprot, simple, lipids : lipidhome.co.uk, Retrieved April 15, 2019

- Lipid-metabolism-signaling-pathway: creative-diagnostics.com, Retrieved July 23, 2019

- Cholest, simple, lipids: lipidhome.co.uk, Retrieved January 29, 2019

- Tag2, simple, lipids: lipidhome.co.uk , Retrieved June 26, 2019

Chapter 4

Protein Metabolism

Proteins are the macromolecules that comprise of one or more than one long chain of amino acid residues. It performs functions like DNA replication, responding to stimuli, catalysing metabolic reactions, provide structure to organisms and cells, transport molecules, etc. The various biomolecule processes responsible for the breakdown of proteins and the synthesis of proteins and amino acids is referred to as protein metabolism. The topics elaborated in this chapter will help in gaining a better perspective about the functions of protein and their protein metabolism.

Protein

Protein is highly complex substance that is present in all living organisms. Proteins are of great nutritional value and are directly involved in the chemical processes essential for life. Proteins are species-specific; that is, the proteins of one species differ from those of another species. They are also organ-specific; for instance, within a single organism, muscle proteins differ from those of the brain and liver.

A protein molecule is very large compared with molecules of sugar or salt and consists of many amino acids joined together to form long chains, much as beads are arranged on a string. There are about 20 different amino acids that occur naturally in proteins. Proteins of similar function have similar amino acid composition and sequence. Although it is not yet possible to explain all of the functions of a protein from its amino acid sequence, established correlations between structure and function can be attributed to the properties of the amino acids that compose proteins.

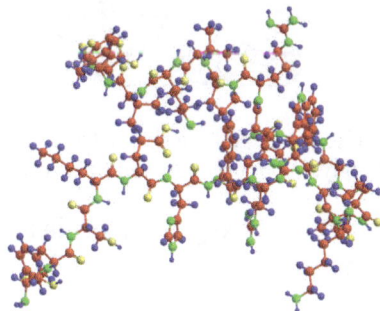

Peptide

The molecular structure of a peptide (a small protein) consists of a sequence of amino acids.

Plants can synthesize all of the amino acids; animals cannot, even though all of them are essential for life. Plants can grow in a medium containing inorganic nutrients that provide nitrogen, potassium, and other substances essential for growth. They utilize the carbon dioxide in the air during the process of photosynthesis to form organic compounds such as carbohydrates. Animals, however, must obtain organic nutrients from outside sources. Because the protein content of most plants is low, very large amounts of plant material are required by animals, such as ruminants (e.g., cows), that eat only plant material to meet their amino acid requirements. Nonruminant animals, including humans, obtain proteins principally from animals and their products—e.g., meat, milk, and eggs. The seeds of legumes are increasingly being used to prepare inexpensive protein-rich food.

The protein content of animal organs is usually much higher than that of the blood plasma. Muscles, for example, contain about 30 percent protein, the liver 20 to 30 percent, and red blood cells 30 percent. Higher percentages of protein are found in hair, bones, and other organs and tissues with a low water content. The quantity of free amino acids and peptides in animals is much smaller than the amount of protein; protein molecules are produced in cells by the stepwise alignment of amino acids and are released into the body fluids only after synthesis is complete.

The high protein content of some organs does not mean that the importance of proteins is related to their amount in an organism or tissue; on the contrary, some of the most important proteins, such as enzymes and hormones, occur in extremely small amounts. The importance of proteins is related principally to their function. All enzymes identified thus far are proteins. Enzymes, which are the catalysts of all metabolic reactions, enable an organism to build up the chemical substances necessary for life—proteins, nucleic acids, carbohydrates, and lipids—to convert them into other substances, and to degrade them. Life without enzymes is not possible. There are several protein hormones with important regulatory functions. In all vertebrates, the respiratory protein hemoglobin acts as oxygen carrier in the blood, transporting oxygen from the lung to body organs and tissues. A large group of structural proteins maintains and protects the structure of the animal body.

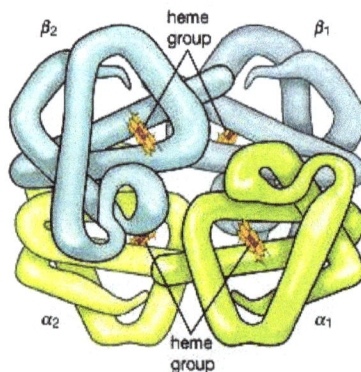

Hemoglobin is a protein made up of four polypeptide chains (α_1, α_2, β_1, and β_2). Each chain is attached to a heme group composed of porphyrin (an organic ringlike compound) attached to an iron atom.

These iron-porphyrin complexes coordinate oxygen molecules reversibly, an ability directly related to the role of hemoglobin in oxygen transport in the blood.

General Structure and Properties of Proteins

Amino Acid Composition of Proteins

The common property of all proteins is that they consist of long chains of α-amino (alpha amino) acids. The general structure of α-amino acids is shown in. The α-amino acids are so called because the α-carbon atom in the molecule carries an amino group ($-NH_2$); the α-carbon atom also carries a carboxyl group ($-COOH$).

In acidic solutions, when the pH is less than 4, the $-COO$ groups combine with hydrogen ions (H^+) and are thus converted into the uncharged form ($-COOH$). In alkaline solutions, at pH above 9, the ammonium groups ($-NH^+_3$) lose a hydrogen ion and are converted into amino groups ($-NH_2$). In the pH range between 4 and 8, amino acids carry both a positive and a negative charge and therefore do not migrate in an electrical field. Such structures have been designated as dipolar ions, or zwitterions (i.e., hybrid ions).

Although more than 100 amino acids occur in nature, particularly in plants, only 20 types are commonly found in most proteins. In protein molecules the α-amino acids are linked to each other by peptide bonds between the amino group of one amino acid and the carboxyl group of its neighbour.

The condensation (joining) of three amino acids yields the tripeptide.

three amino acids joined by peptide bonds

It is customary to write the structure of peptides in such a way that the free α-amino group (also called the N terminus of the peptide) is at the left side and the free carboxyl group (the Cterminus) at the right side. Proteins are macromolecular polypeptides—i.e., very large molecules composed of many peptide-bonded amino acids. Most of the common ones contain more than 100 amino acids linked to each other in a long peptide chain. The average molecular weight (based on the weight of a hydrogen atom as 1) of each amino acid is approximately 100 to 125; thus, the molecular weights of proteins are usually in the range of 10,000 to 100,000 daltons (one dalton is the weight of one hydrogen atom). The species-specificity and organ-specificity of proteins result from differences in the number and sequences of amino acids. Twenty different amino acids in a chain 100 amino acids long can be arranged in far more than 10^{100} ways (10^{100} is the number one followed by 100 zeroes).

Structures of Common Amino Acids

The amino acids present in proteins differ from each other in the structure of their side (R) chains.

The simplest amino acid is glycine, in which R is a hydrogen atom. In a number of amino acids, R represents straight or branched carbon chains. One of these amino acids is alanine, in which R is the methyl group ($-CH_3$). Valine, leucine, and isoleucine, with longer R groups, complete the alkyl side-chain series. The alkyl side chains (R groups) of these amino acids are nonpolar; this means that they have no affinity for water but some affinity for each other. Although plants can form all of the alkyl amino acids, animals can synthesize only alanine and glycine; thus valine, leucine, and isoleucine must be supplied in the diet.

Two amino acids, each containing three carbon atoms, are derived from alanine; they are serine and cysteine. Serine contains an alcohol group ($-CH_2OH$) instead of the methyl group of alanine, and cysteine contains a mercapto group ($-CH_2SH$). Animals can synthesize serine but not cysteine or cystine. Cysteine occurs in proteins predominantly in its oxidized form (oxidation in this sense meaning the removal of hydrogen atoms), called cystine. Cystine consists of two cysteine molecules linked by the disulfide bond ($-S-S-$) that results when a hydrogen atom is removed from the mercapto group of each of the cysteines. Disulfide bonds are important in protein structure because they allow the linkage of two different parts of a protein molecule to—and thus the formation of loops in—the otherwise straight chains. Some proteins contain small amounts of cysteine with free sulfhydryl ($-SH$) groups.

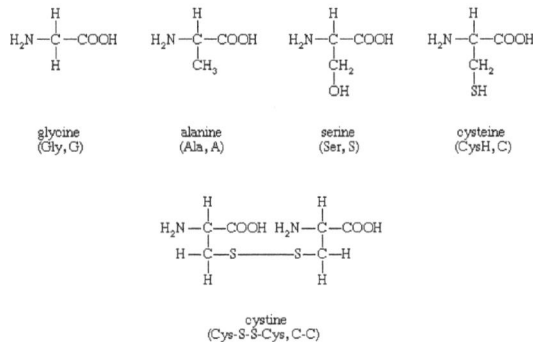

glycine (Gly, G)   alanine (Ala, A)   serine (Ser, S)   cysteine (CysH, C)

cystine (Cys-S-S-Cys, C-C)

Four amino acids, each consisting of four carbon atoms, occur in proteins; they are aspartic acid, asparagine, threonine, and methionine. Aspartic acid and asparagine, which occur in large amounts, can be synthesized by animals. Threonine and methionine cannot be synthesized and thus are essential amino acids; i.e., they must be supplied in the diet. Most proteins contain only small amounts of methionine.

Proteins also contain an amino acid with five carbon atoms (glutamic acid) and a secondary amine (in proline), which is a structure with the amino group ($-NH_2$) bonded to the alkyl side chain, forming a ring. Glutamic acid and aspartic acid are dicarboxylic acids; that is, they have two carboxyl groups ($-COOH$).

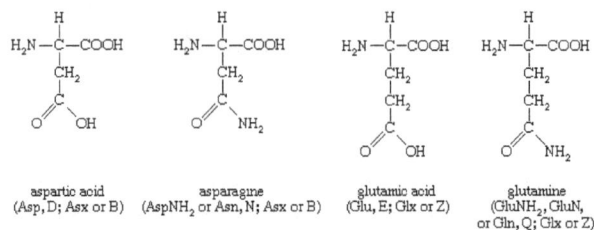

aspartic acid (Asp, D; Asx or B)   asparagine (AspNH₂ or Asn, N; Asx or B)   glutamic acid (Glu, E; Glx or Z)   glutamine (GluNH₂, GluN, or Gln, Q; Glx or Z)

Glutamine is similar to asparagine in that both are the amides of their corresponding dicarboxylic acid forms; i.e., they have an amide group ($-CONH_2$) in place of the carboxyl ($-COOH$) of the side chain. Glutamic acid and glutamine are abundant in most proteins; e.g., in plant proteins they sometimes comprise more than one-third of the amino acids present. Both glutamic acid and glutamine can be synthesized by animals.

Amino Acid Content of Some Proteins

amino acid	protein					
	alpha-casein	gliadin	edestin	collagen (ox hide)	keratin (wool)	myosin
lysine	60.9	4.45	19.9	27.4	6.2	85
histidine	18.7	11.7	18.6	4.5	19.7	15
arginine	24.7	15.7	99.2	47.1	56.9	41
aspartic acid**	63.1	10.1	99.4	51.9	51.5	85
threonine	41.2	17.6	31.2	19.3	55.9	41
serine	63.1	46.7	55.7	41.0	79.5	41
glutamic acid**	153.1	311.0	144.9	76.2	99.0	155
proline	71.3	117.8	32.9	125.2	58.3	22
glycine	37.3	—	68.0	354.6	78.0	39
alanine	41.5	23.9	57.7	115.7	43.8	78
half-cystine	3.6	21.3	10.9	0.0	105.0	86
valine	53.8	22.7	54.6	21.4	46.6	42
methionine	16.8	11.3	16.4	6.5	4.0	22
isoleucine	48.8	90.8***	41.9	14.5	29.0	42
leucine	60.3		60.0	28.2	59.9	79
tyrosine	44.7	17.7	26.9	5.5	28.7	18
phenylalanine	27.9	39.0	38.4	13.9	22.4	27
tryptophan	7.8	3.2	6.6	0.0	9.6	—
hydroxyproline	0.0	0.0	0.0	97.5	12.2	—
hydroxylysine	—	—	—	8.0	1.2	—
total	839	765	883	1,058	863	832
average residual weight	119	131	113	95	117	120

The amino acids proline and hydroxyproline occur in large amounts in collagen, the protein of the connective tissue of animals. Proline and hydroxyproline lack free amino ($-NH_2$) groups because the amino group is enclosed in a ring structure with the side chain; they thus cannot exist in a zwitterion form. Although the nitrogen-containing group (>NH) of these amino acids can form a peptide bond with the carboxyl group of another amino acid, the bond so formed gives rise to a kink in the peptide chain; i.e., the ring structure alters the regular bond angle of normal peptide bonds.

Proteins usually are almost neutral molecules; that is, they have neither acidic nor basic properties. This means that the acidic carboxyl ($-COO^-$) groups of aspartic and glutamic acid are about equal in number to the amino acids with basic side chains. Three such basic amino acids, each containing six carbon atoms, occur in proteins. The one with the simplest structure, lysine, is synthesized by plants but not by animals. Even some plants have a low lysine content. Arginine is found in all proteins; it occurs in particularly high amounts in the strongly basic protamines (simple proteins composed of relatively few amino acids) of fish sperm. The third basic amino acid is histidine. Both arginine and histidine can be synthesized by animals. Histidine is a weaker base than either lysine or arginine. The imidazole ring, a five-membered ring structure containing two nitrogen atoms in the side chain of histidine, acts as a buffer (i.e., a stabilizer of hydrogen ion concentration) by binding hydrogen ions (H^+) to the nitrogen atoms of the imidazole ring.

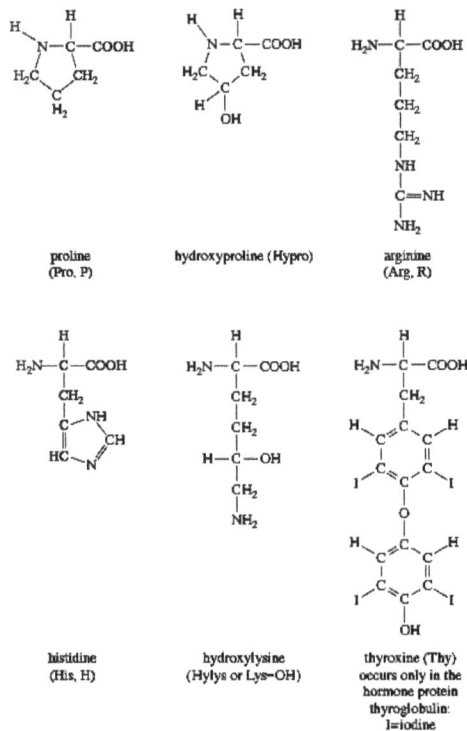

proline (Pro, P)　　　hydroxyproline (Hypro)　　　arginine (Arg, R)

histidine (His, H)　　　hydroxylysine (Hylys or Lys—OH)　　　thyroxine (Thy) occurs only in the hormone protein thyroglobulin: I=iodine

The remaining amino acids—phenylalanine, tyrosine, and tryptophan—have in common an aromatic structure; i.e., a benzene ring is present. These three amino acids are essential, and, while animals cannot synthesize the benzene ring itself, they can convert phenylalanine to tyrosine.

Because these amino acids contain benzene rings, they can absorb ultraviolet light at wavelengths between 270 and 290 nanometres (nm; 1 nanometre = 10^{-9} metre = 10 angstrom units). Phenylalanine absorbs very little ultraviolet light; tyrosine and tryptophan, however, absorb it strongly and are responsible for the absorption band most proteins exhibit at 280–290 nanometres. This absorption is often used to determine the quantity of protein present in protein samples.

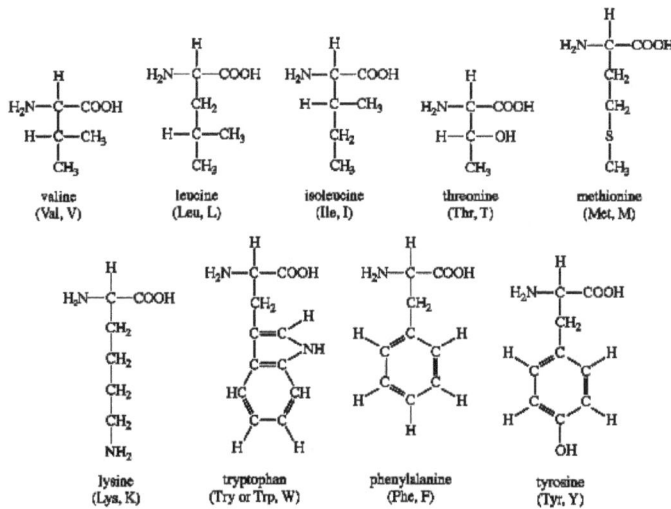

valine (Val, V); leucine (Leu, L); isoleucine (Ile, I); threonine (Thr, T); methionine (Met, M); lysine (Lys, K); tryptophan (Try or Trp, W); phenylalanine (Phe, F); tyrosine (Tyr, Y)

Most proteins contain only the amino acids described above; however, other amino acids occur in proteins in small amounts. For example, the collagen found in connective tissue contains, in addition to hydroxyproline, small amounts of hydroxylysine. Other proteins contain some monomethyl-, dimethyl-, or trimethyllysine—i.e., lysine derivatives containing one, two, or three methyl groups ($-CH_3$). The amount of these unusual amino acids in proteins, however, rarely exceeds 1 or 2 percent of the total amino acids.

Physicochemical Properties of the Amino Acids

The physicochemical properties of a protein are determined by the analogous properties of the amino acids in it.

The α-carbon atom of all amino acids, with the exception of glycine, is asymmetric; this means that four different chemical entities (atoms or groups of atoms) are attached to it. As a result, each of the amino acids, except glycine, can exist in two different spatial, or geometric, arrangements (i.e., isomers), which are mirror images akin to right and left hands.

These isomers exhibit the property of optical rotation. Optical rotation is the rotation of the plane of polarized light, which is composed of light waves that vibrate in one plane, or direction, only. Solutions of substances that rotate the plane of polarization are said to be optically active, and the degree of rotation is called the optical rotation of the solution. The direction in which the light is rotated is generally designed as plus, or d, for dextrorotatory (to the right), or as minus, or l, for levorotatory (to the left). Some amino acids are dextrorotatory, others are levorotatory. With the exception of a few small proteins (peptides) that occur in bacteria, the amino acids that occur in proteins are L-amino acids.

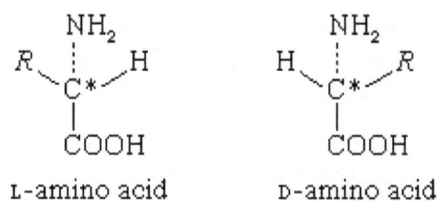

L-amino acid   D-amino acid

In bacteria, D-alanine and some other D-amino acids have been found as components of gramicidin and bacitracin. These peptides are toxic to other bacteria and are used in medicine as antibiotics. The D-alanine has also been found in some peptides of bacterial membranes.

In contrast to most organic acids and amines, the amino acids are insoluble in organic solvents. In aqueous solutions they are dipolar ions (zwitterions, or hybrid ions) that react with strong acids or bases in a way that leads to the neutralization of the negatively or positively charged ends, respectively. Because of their reactions with strong acids and strong bases, the amino acids act as buffers—stabilizers of hydrogen ion (H^+) or hydroxide ion (OH^-) concentrations. In fact, glycine is frequently used as a buffer in the pH range from 1 to 3 (acid solutions) and from 9 to 12 (basic solutions). In acid solutions, glycine has a positive charge and therefore migrates to the cathode (negative electrode of a direct-current electrical circuit with terminals in the solution). Its charge, however, is negative in alkaline solutions, in which it migrates to the anode (positive electrode). At pH 6.1 glycine does not migrate, because each molecule has one positive and one negative charge. The pH at which an amino acid does not migrate in an electrical field is called the isoelectric point. Most of the monoamino acids (i.e., those with only one amino group) have isoelectric points similar to that of glycine. The isoelectric points of aspartic and glutamic acids, however, are close to pH 3, and those of histidine, lysine, and arginine are at pH 7.6, 9.7, and 10.8, respectively.

Amino Acid Sequence in Protein Molecules

Since each protein molecule consists of a long chain of amino acid residues, linked to each other by peptide bonds, the hydrolytic cleavage of all peptide bonds is a prerequisite for the quantitative determination of the amino acid residues. Hydrolysis is most frequently accomplished by boiling the protein with concentrated hydrochloric acid. The quantitative determination of the amino acids is based on the discovery that amino acids can be separated from each other by chromatography on filter paper and made visible by spraying the paper with ninhydrin. The amino acids of the protein hydrolysate are separated from each other by passing the hydrolysate through a column of adsorbents, which adsorb the amino acids with different affinities and, on washing the column with buffer solutions, release them in a definite order. The amount of each of the amino acids can be determined by the intensity of the colour reaction with ninhydrin.

To obtain information about the sequence of the amino acid residues in the protein, the protein is degraded stepwise, one amino acid being split off in each step. This is accomplished by coupling the free α-amino group ($-NH_2$) of the N-terminal amino acid with phenyl isothiocyanate; subsequent mild hydrolysis does not affect the peptide bonds. The procedure, called the Edman degradation, can be applied repeatedly; it thus reveals the sequence of the amino acids in the peptide chain.

Unavoidable small losses that occur during each step make it impossible to determine the sequence of more than about 30 to 50 amino acids by this procedure. For this reason the protein is usually first hydrolyzed by exposure to the enzyme trypsin, which cleaves only peptide bonds formed by the carboxyl groups of lysine and arginine. The Edman degradationis then applied to each of the few resulting peptides produced by the action of trypsin.

Levels of Structural Organization in Proteins

Primary Structure

Analytical and synthetic procedures reveal only the primary structure of the proteins—that is, the amino acid sequence of the peptide chains. They do not reveal information about the conformation (arrangement in space) of the peptide chain—that is, whether the peptide chain is present as a long straight thread or is irregularly coiled and folded into a globule. The configuration, or conformation, of a protein is determined by mutual attraction or repulsion of polar or nonpolar groups in the side chains (R groups) of the amino acids. The former have positive or negative charges in their side chains; the latter repel water but attract each other. Some parts of a peptide chain containing 100 to 200 amino acids may form a loop, or helix; others may be straight or form irregular coils.

The terms *secondary*, *tertiary*, and *quaternary structure* are frequently applied to the configuration of the peptide chain of a protein. A nomenclature committee of the International Union of Biochemistry (IUB) has defined these terms as follows: The primary structure of a protein is determined by its amino acid sequence without any regard for the arrangement of the peptide chain in space. The secondary structure is determined by the spatial arrangement of the main peptide chain without any regard for the conformation of side chains or other segments of the main chain. The tertiary structure is determined by both the side chains and other adjacent segments of the main chain, without regard for neighbouring peptide chains. Finally, the term *quaternary structure* is used for the arrangement of identical or different subunits of a large protein in which each subunit is a separate peptide chain.

Secondary Structure

The nitrogen and carbon atoms of a peptide chain cannot lie on a straight line, because of the magnitude of the bond angles between adjacent atoms of the chain; the bond angle is about 110°. Each of the nitrogen and carbon atoms can rotate to a certain extent, however, so that the chain has a limited flexibility. Because all of the amino acids, except glycine, are asymmetric L-amino acids, the peptide chain tends to assume an asymmetric helical shape; some of the fibrous proteins consist of elongated helices around a straight screw axis. Such structural features result from properties common to all peptide chains. The product of their effects is the secondary structure of the protein.

Tertiary Structure

The tertiary structure is the product of the interaction between the side chains (R) of the amino acids composing the protein. Some of them contain positively or negatively charged groups, others are polar, and still others are nonpolar. The number of carbon atoms in the side chain varies from zero in glycine to nine in tryptophan. Positively and negatively charged side chains have the tendency to attract each other; side chains with identical charges repel each other. The bonds formed by the forces between the negatively charged side chains of aspartic or glutamic acid on the one hand, and the positively charged side chains of lysine or arginine on the other hand, are called salt bridges. Mutual attraction of adjacent peptide chains also results from the formation of numerous hydrogen bonds.

direction
of C terminal

peptide
chains

—hydrogen bonds

direction
of C terminal

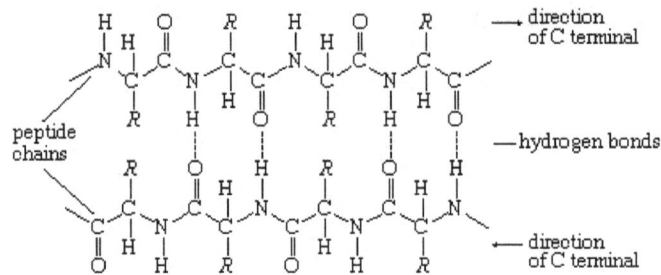

Hydrogen bonds form as a result of the attraction between the nitrogen-bound hydrogen atom (the imide hydrogen) and the unshared pair of electrons of the oxygen atom in the double bonded carbon–oxygen group (the carbonyl group). The result is a slight displacement of the imide hydrogen toward the oxygen atom of the carbonyl group. Although the hydrogen bond is much weaker than a covalent bond (i.e., the type of bond between two carbon atoms, which equally share the pair of bonding electrons between them), the large number of imide and carbonyl groups in peptide chains results in the formation of numerous hydrogen bonds. Another type of attraction is that between nonpolar side chains of valine, leucine, isoleucine, and phenylalanine; the attraction results in the displacement of water molecules and is called hydrophobic interaction.

In proteins rich in cystine, the conformation of the peptide chain is determined to a considerable extent by the disulfide bonds (−S−S−) of cystine. The halves of cystine may be located in different parts of the peptide chain and thus may form a loop closed by the disulfide bond.

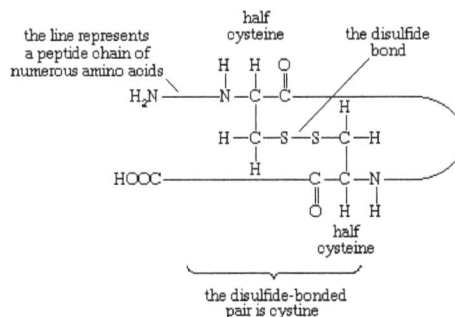

the line represents
a peptide chain of
numerous amino acids

half
cysteine

the disulfide
bond

half
cysteine

the disulfide-bonded
pair is cystine

If the disulfide bond is reduced (i.e., hydrogen is added) to two sulfhydryl (−SH) groups, the tertiary structure of the protein undergoes a drastic change—closed loops are broken and adjacent disulfide-bonded peptide chains separate.

Quaternary Structure

The nature of the quaternary structure is demonstrated by the structure of hemoglobin. Each molecule of human hemoglobin consists of four peptide chains, two α-chains and two β-chains; i.e., it is a tetramer. The four subunits are linked to each other by hydrogen bonds and hydrophobic interaction. Because the four subunits are so closely linked, the hemoglobin tetramer is called a molecule, even though no covalent bonds occur between the peptide chains of the four subunits. In other proteins, the subunits are bound to each other by covalent bonds (disulfide bridges).

The amino acid sequence of porcine proinsulin is shown below. The arrows indicate the direction from the N terminus of the β-chain (B) to the C terminus of the α-chain (A).

F.V.N.Q.H.L.C.G.S.H.L.V.E.A.L.Y.L.V.C.G.E.R.G.F.F.Y.T.P.K.A   *B* chain ⎫
\|—— disulfide bonds ——\   ⎬ insulin
G.I.V.E.Q.C.C.T.S.I.C.S.L.Y.Q.L.E.N.Y.C.N   *A* chain ⎭
      ⌊——⌋ —— disulfide bond
R.K.Q.P.P.G.E.L.A.L.A.Q.L.G.G.L.G.G.G.L.E.V.A.G.A.Q.P.N.Q.A.E.A.A   *C* peptide

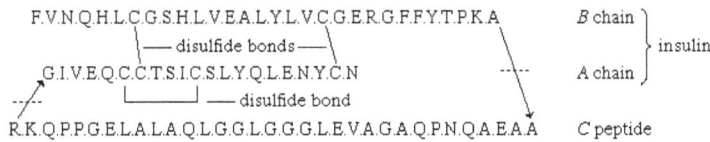

Isolation and Determination of Proteins

Animal material usually contains large amounts of protein and lipids and small amounts of carbohydrate; in plants, the bulk of the dry matter is usually carbohydrate. If it is necessary to determine the amount of protein in a mixture of animal foodstuffs, a sample is converted to ammonium salts by boiling with sulfuric acid and a suitable inorganic catalyst, such as copper sulfate (Kjeldahl method). The method is based on the assumption that proteins contain 16 percent nitrogen, and that nonprotein nitrogen is present in very small amounts. The assumption is justified for most tissues from higher animals but not for insects and crustaceans, in which a considerable portion of the body nitrogen is present in the form of chitin, a carbohydrate. Large amounts of nonprotein nitrogen are also found in the sap of many plants. In such cases, the precise quantitative analyses are made after the proteins have been separated from other biological compounds.

Proteins are sensitive to heat, acids, bases, organic solvents, and radiation exposure; for this reason, the chemical methods employed to purify organic compounds cannot be applied to proteins. Salts and molecules of small size are removed from protein solutions by dialysis—i.e., by placing the solution into a sac of semipermeable material, such as cellulose or acetylcellulose, which will allow small molecules to pass through but not large protein molecules, and immersing the sac in water or a salt solution. Small molecules can also be removed either by passing the protein solution through a column of resin that adsorbs only the protein or by gel filtration. In gel filtration, the large protein molecules pass through the column, and the small molecules are adsorbed to the gel.

Groups of proteins are separated from each other by salting out—i.e., the stepwise addition of sodium sulfate or ammonium sulfate to a protein solution. Some proteins, called globulins, become insoluble and precipitate when the solution is half-saturated with ammonium sulfate or when its sodium sulfate content exceeds about 12 percent. Other proteins, the albumins, can be precipitated from the supernatant solution (i.e., the solution remaining after a precipitation has taken place) by saturation with ammonium sulfate. Water-soluble proteins can be obtained in a dry state by freeze-drying (lyophilization), in which the protein solution is deep-frozen by lowering the temperature below −15 °C (5 °F) and removing the water; the protein is obtained as a dry powder.

Most proteins are insoluble in boiling water and are denatured by it—i.e., irreversibly converted into an insoluble material. Heat denaturation cannot be used with connective tissue because the principal structural protein, collagen, is converted by boiling water into water-soluble gelatin.

Fractionation (separation into components) of a mixture of proteins of different molecular weight can be accomplished by gel filtration. The size of the proteins retained by the gel depends upon the properties of the gel. The proteins retained in the gel are removed from the column by solutions of a suitable concentration of salts and hydrogen ions.

Many proteins were originally obtained in crystalline form, but crystallinity is not proof of purity; many crystalline protein preparations contain other substances. Various tests are used to determine whether a protein preparation contains only one protein. The purity of a protein solution can be determined by such techniques as chromatography and gel filtration. In addition, a solution of pure protein will yield one peak when spun in a centrifuge at very high speeds (ultracentrifugation) and will migrate as a single band in electrophoresis (migration of the protein in an electrical field). After these methods and others (such as amino acid analysis) indicate that the protein solution is pure, it can be considered so. Because chromatography, ultracentrifugation, and electrophoresis cannot be applied to insoluble proteins, little is known about them; they may be mixtures of many similar proteins.

Very small (microheterogeneous) differences in some of the apparently pure proteins are known to occur. They are differences in the amino acid composition of otherwise identical proteins and are transmitted from generation to generation; i.e., they are genetically determined. For example, some humans have two hemoglobins, hemoglobin A and hemoglobin S, which differ in one amino acid at a specific site in the molecule. In hemoglobin A the site is occupied by glutamic acid and in hemoglobin S by valine. Refinement of the techniques of protein analysis has resulted in the discovery of other instances of microheterogeneity.

The quantity of a pure protein can be determined by weighing or by measuring the ultraviolet absorbancy at 280 nanometres. The absorbency at 280 nanometres depends on the content of tyrosine and tryptophan in the protein. Sometimes the slightly less sensitive biuret reaction, a purple colour given by alkaline protein solutions upon the addition of copper sulfate, is used; its intensity depends only on the number of peptide bonds per gram, which is similar in all proteins.

Physicochemical Properties of Proteins

Molecular Weight of Proteins

The molecular weight of proteins cannot be determined by the methods of classical chemistry (e.g., freezing-point depression), because they require solutions of a higher concentration of protein than can be prepared.

If a protein contains only one molecule of one of the amino acids or one atom of iron, copper, or another element, the minimum molecular weight of the protein or a subunit can be calculated; for example, the protein myoglobin contains 0.34 gram of iron in 100 grams of protein. The atomic weight of iron is 56; thus the minimum molecular weight of myoglobin is $(56 \times 100)/0.34$ = about 16,500. Direct measurements of the molecular weight of myoglobin yield the same value. The molecular weight of hemoglobin, however, which also contains 0.34 percent iron, has been found to be 66,000 or $4 \times 16,500$; thus hemoglobin contains four atoms of iron.

The method most frequently used to determine the molecular weight of proteins is ultracentrifugation—i.e., spinning in a centrifuge at velocities up to about 60,000 revolutions per minute. Centrifugal forces of more than 200,000 times the gravitational force on the surface of Earth are achieved at such velocities. The first ultracentrifuges, built in 1920, were used to determine the molecular weight of proteins. The molecular weights of a large number of proteins have been determined. Most consist of several subunits, the molecular weight of which is usually less than

100,000 and frequently ranges from 20,000 to 30,000. Proteins of very high molecular weights are found among hemocyanins, the copper-containing respiratory proteins of invertebrates; some range as high as several million. Although there is no definite lower limit for the molecular weight of proteins, short amino acid sequences are usually called peptides.

Shape of Protein Molecules

In the technique of X-ray diffraction, the X-rays are allowed to strike a protein crystal. The X-rays, diffracted (bent) by the crystal, impinge on a photographic plate, forming a pattern of spots. This method reveals that peptide chains can assume very complicated, apparently irregular shapes. Two extremes in shape include the closely folded structure of the globular proteins and the elongated, unidimensional structure of the threadlike fibrous proteins; both were recognized many years before the technique of X-ray diffraction was developed. Solutions of fibrous proteins are extremely viscous (i.e., sticky); those of the globular proteins have low viscosity (i.e., they flow easily). A 5 percent solution of a globular protein—ovalbumin, for example—easily flows through a narrow glass tube; a 5 percent solution of gelatin, a fibrous protein, however, does not flow through the tube, because it is liquid only at high temperatures and solidifies at room temperature. Even solutions containing only 1 or 2 percent of gelatin are highly viscous and flow through a narrow tube either very slowly or only under pressure.

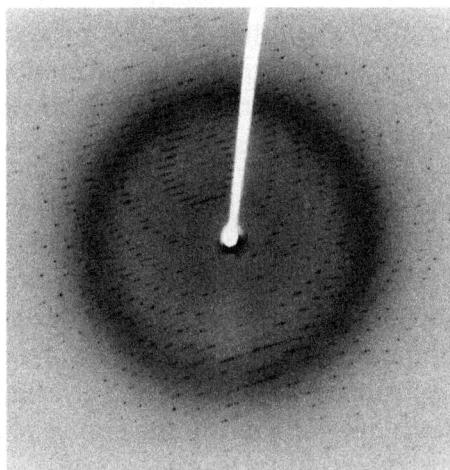

X-ray diffraction
X-ray diffraction pattern of a crystallized enzyme.

The elongated peptide chains of the fibrous proteins can be imagined to become entangled not only mechanically but also by mutual attraction of their side chains, and in this way they incorporate large amounts of water. Most of the hydrophilic (water-attracting) groups of the globular proteins, however, lie on the surface of the molecules, and, as a result, globular proteins incorporate only a few water molecules. If a solution of a fibrous protein flows through a narrow tube, the elongated molecules become oriented parallel to the direction of the flow, and the solution thus becomes birefringent like a crystal; i.e., it splits a light ray into two components that travel at different velocities and are polarized at right angles to each other. Globular proteins do not show this phenomenon, which is called flow birefringence. Solutions of myosin, the contractile protein of muscles, show very high flow birefringence; other proteins with very high flow birefringence include solutions of fibrinogen, the clotting material of blood plasma, and solutions of tobacco mosaic virus.

The gamma-globulins of the blood plasma show low flow birefringence, and none can be observed in solutions of serum albumin and ovalbumin.

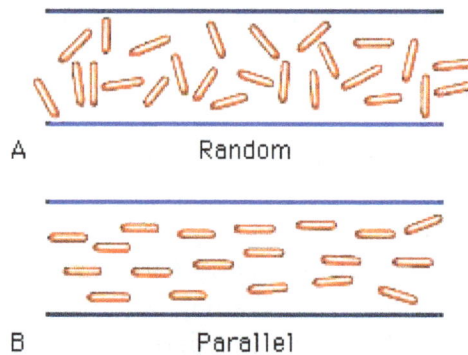

Figure: Flow birefringence. Orientation of elongated, rodlike macromolecules
(A) in resting solution, or (B) during flow through a horizontal tube.

Hydration of Proteins

When dry proteins are exposed to air of high water content, they rapidly bind water up to a maximum quantity, which differs for different proteins; usually it is 10 to 20 percent of the weight of the protein. The hydrophilic groups of a protein are chiefly the positively charged groups in the side chains of lysine and arginine and the negatively charged groups of aspartic and glutamic acid. Hydration (i.e., the binding of water) may also occur at the hydroxyl ($-OH$) groups of serine and threonine or at the amide ($-CONH_2$) groups of asparagine and glutamine.

The binding of water molecules to either charged or polar (partly charged) groups is explained by the dipolar structure of the water molecule; that is, the two positively charged hydrogen atoms form an angle of about 105°, with the negatively charged oxygen atom at the apex. The centre of the positive charges is located between the two hydrogen atoms; the centre of the negative charge of the oxygen atom is at the apex of the angle. The negative pole of the dipolar water molecule binds to positively charged groups; the positive pole binds negatively charged ones. The negative pole of the water molecule also binds to the hydroxyl and amino groups of the protein.

The water of hydration is essential to the structure of protein crystals; when they are completely dehydrated, the crystalline structure disintegrates. In some proteins this process is accompanied by denaturation and loss of the biological function.

In aqueous solutions, proteins bind some of the water molecules very firmly; others are either very loosely bound or form islands of water molecules between loops of folded peptide chains. Because the water molecules in such an island are thought to be oriented as in ice, which is crystalline water, the islands of water in proteins are called icebergs. Water molecules may also form bridges between the carbonyl and imino groups of adjacent peptide chains, resulting in structures similar to those of the pleated sheet but with a water molecule in the position of the hydrogen bonds of that configuration. The extent of hydration of protein molecules in aqueous solutions is important, because some of the methods used to determine the molecular weight of proteins yield the molecular weight of the hydrated protein. The amount of water bound to one gram of a globular protein in solution varies from 0.2 to 0.5 gram. Much larger amounts of water are mechanically immobilized between the elongated peptide chains of fibrous proteins; for example, one gram of gelatin can immobilize at room temperature 25 to 30 grams of water.

Hydration of proteins is necessary for their solubility in water. If the water of hydration of a protein dissolved in water is reduced by the addition of a salt such as ammonium sulfate, the protein is no longer soluble and is salted out, or precipitated. The salting-out process is reversible because the protein is not denatured (i.e., irreversibly converted to an insoluble material) by the addition of such salts as sodium chloride, sodium sulfate, or ammonium sulfate. Some globulins, called euglobulins, are insoluble in water in the absence of salts; their insolubility is attributed to the mutual interaction of polar groups on the surface of adjacent molecules, a process that results in the formation of large aggregates of molecules. Addition of small amounts of salt causes the euglobulins to become soluble. This process, called salting in, results from a combination between anions (negatively charged ions) and cations (positively charged ions) of the salt and positively and negatively charged side chains of the euglobulins. The combination prevents the aggregation of euglobulin molecules by preventing the formation of salt bridges between them. The addition of more sodium or ammonium sulfate causes the euglobulins to salt out again and to precipitate.

Electrochemistry of Proteins

Because the α-amino group and α-carboxyl group of amino acids are converted into peptide bonds in the protein molecule, there is only one α-amino group (at the N terminus) and one α-carboxyl group (at the C terminus) in a given protein molecule. The electrochemical character of a protein is affected very little by these two groups. Of importance, however, are the numerous positively charged ammonium groups ($-NH_3^+$) of lysine and arginine and the negatively charged carboxyl groups ($-COO^-$) of aspartic acid and glutamic acid. In most proteins, the number of positively and negatively charged groups varies from 10 to 20 per 100 amino acids.

Electrometric Titration

When measured volumes of hydrochloric acid are added to a solution of protein in salt-free water, the pH decreases in proportion to the amount of hydrogen ions added until it is about 4. Further addition of acid causes much less decrease in pH because the protein acts as a buffer at pH values of 3 to 4. The reaction that takes place in this pH range is the protonation of the carboxyl group—i.e., the conversion of $-COO-$ into $-COOH$. Electrometric titration of an isoelectric protein with potassium hydroxide causes a very slow increase in pH and a weak buffering action of the protein at pH 7; a very strong buffering action occurs in the pH range from 9 to 10. The buffering action at pH 7, which is caused by loss of protons (positively charged hydrogen) from the imidazolium groups (i.e., the five-member ring structure in the side chain) of histidine, is weak because the histidine content of proteins is usually low. The much stronger buffering action at pH values from 9 to 10 is caused by the loss of protons from the hydroxyl group of tyrosine and from the ammonium groups of lysine. Finally, protons are lost from the guanidinium groups (i.e., the nitrogen-containing terminal portion of the arginine side chains) of arginine at pH 12. Electrometric titrations of proteins yield similar curves. Electrometric titration makes possible the determination of the approximate number of carboxyl groups, ammonium groups, histidines, and tyrosines per molecule of protein.

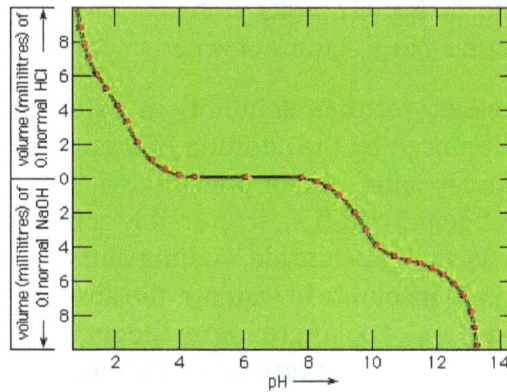

Figure: Electrometric titration of glycine.

Electrophoresis

The positively and negatively charged side chains of proteins cause them to behave like amino acids in an electrical field; that is, they migrate during electrophoresis at low pH values to the cathode (negative terminal) and at high pH values to the anode (positive terminal). The isoelectric point, the pH value at which the protein molecule does not migrate, is in the range of pH 5 to 7 for many proteins. Proteins such as lysozyme, cytochrome c, histone, and others rich in lysine and arginine, however, have isoelectric points in the pH range between 8 and 10. The isoelectric point of pepsin, which contains very few basic amino acids, is close to 1.

Number of Amino Acids Per Protein Molecule

Amino acid	Protein						
	Cyto	Hb alpha	Hb beta	RNase	Lys	Chgen	Fdox
Lysine	18	11	11	10	6	14	4
Histidine	3	10	9	4	1	2	1
Arginine	2	3	3	4	11	4	1
Aspartic acid**	8	12	13	15	21	23	13
Threonine	7	9	7	10	7	23	8
Serine	2	11	5	15	10	28	7
Glutamic acid**	10	5	11	12	5	15	13
Proline	4	7	7	4	2	9	4
Glycine	13	7	13	3	12	23	6
Alanine	6	21	15	12	12	22	9
Half-cystine	2	1	2	8	8	10	5
Valine	3	13	18	9	6	23	7
Methionine	3	2	1	4	2	2	0

Isoleucine	8	0	0	3	6	10	4
Leucine	6	18	18	2	8	19	8
Tyrosine	5	3	3	6	3	4	4
Phenylalanine	3	7	8	3	3	6	2
Tryptophan	1	1	2	0	6	8	1
Total	104	141	146	124	129	245	97

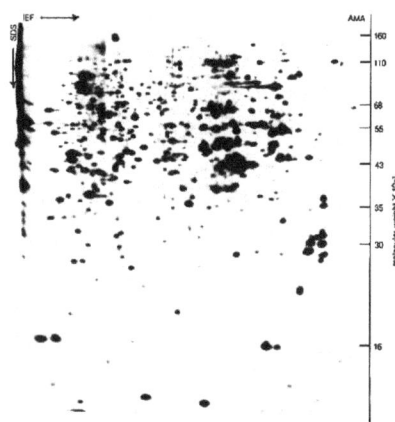

Two-dimensional gel electrophoresis

In two-dimensional gel electrophoresis, proteins are separated based on charge and size. Approaches commonly employed include isoelectric focusing (IEF) sodium dodecyl sulfate (SDS) polyacrylamide gel electrophoresis (PAGE) and immobilized pH gradient (IPG-Dalt) SDS-PAGE.

Free-boundary electrophoresis, the original method of determining electrophoretic migration, has been replaced in many instances by zone electrophoresis, in which the protein is placed in either a gel of starch, agar, or polyacrylamide or in a porous medium such as paper or cellulose acetate. The migration of hemoglobin and other coloured proteins can be followed visually. Colourless proteins are made visible after the completion of electrophoresis by staining them with a suitable dye.

Conformation of Globular Proteins

Results of X-ray diffraction Studies

Most knowledge concerning secondary and tertiary structure of globular proteins has been obtained by the examination of their crystals using X-ray diffraction. In this technique, X-rays are allowed to strike the crystal; the X-rays are diffracted by the crystal and impinge on a photographic plate, forming a pattern of spots. The measured intensity of the diffraction pattern, as recorded on a photographic film, depends particularly on the electron density of the atoms in the protein crystal. This density is lowest in hydrogen atoms, and they do not give a visible diffraction pattern. Although carbon, oxygen, and nitrogen atoms yield visible diffraction patterns, they are present in such great number—about 700 or 800 per 100 amino acids—that the resolution of the structure of a protein containing more than 100 amino acids is almost impossible. Resolution is considerably improved by substituting into the side chains of certain amino acids very heavy atoms, particularly those of heavy metals. Mercury ions, for example, bind to the sulfhydryl (−SH) groups of cysteine.

Platinum chloride has been used in other proteins. In the iron-containing proteins, the iron atom already in the molecule is adequate.

Although the X-ray diffraction technique cannot resolve the complete three-dimensional conformation (that is, the secondary and tertiary structure of the peptide chain), complete resolution has been obtained by combination of the results of X-ray diffraction with those of amino acid sequence analysis. In this way the complete conformation of such proteins as myoglobin, chymotrypsinogen, lysozyme, and ribonuclease has been resolved.

The X-ray diffraction method has revealed regular structural arrangements in proteins; one is an extended form of antiparallel peptide chains that are linked to each other by hydrogen bonds between the carbonyl and imino groups. This conformation, called the pleated sheet, or β-structure, is found in some fibrous proteins. Short strands of the β-structure have also been detected in some globular proteins.

A second important structural arrangement is the α-helix; it is formed by a sequence of amino acids wound around a straight axis in either a right-handed or a left-handed spiral. Each turn of the helix corresponds to a distance of 5.4 angstroms (= 0.54 nanometre) in the direction of the screw axis and contains 3.7 amino acids. Hence, the length of the α-helix per amino acid residue is 5.4 divided by 3.7, or 1.5 angstroms (1 angstrom = 0.1 nanometre). The stability of the α-helix is maintained by hydrogen bonds between the carbonyl and imino groups of neighbouring turns of the helix. It was once thought, based on data from analyses of the myoglobin molecule, more than half of which consists of α-helices, that the α-helix is the predominant structural element of the globular proteins; it is now known that myoglobin is exceptional in this respect. The other globular proteins for which the structures have been resolved by X-ray diffraction contain only small regions of α-helix. In most of them the peptide chains are folded in an apparently random fashion, for which the term random coil has been used. The term is misleading, however, because the folding is not random; rather, it is dictated by the primary structure and modified by the secondary and tertiary structures.

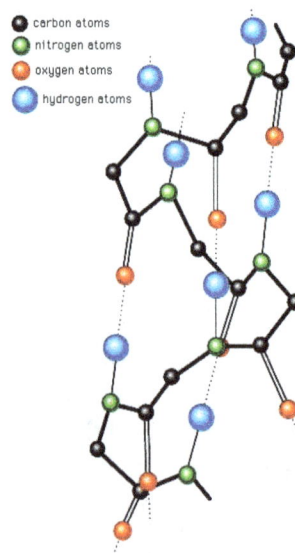

Protein structure; α-helix
The α-helix in the structural arrangement of a protein.

The first proteins for which the internal structures were completely resolved are the iron-containing proteins myoglobin and hemoglobin. The investigation of the hydrated crystals of these proteins by Austrian-born British biochemist Max Perutz and British biochemist John C. Kendrew, who won the 1962 Nobel Prize for Chemistry for their work, revealed that the folding of the peptide chains is so tight that most of the water is displaced from the centre of the globular molecules. The amino acids that carry the ammonium ($-NH_3^+$) and carboxyl ($-COO^-$) groups were found to be shifted to the surface of the globular molecules, and the nonpolar amino acids were found to be concentrated in the interior.

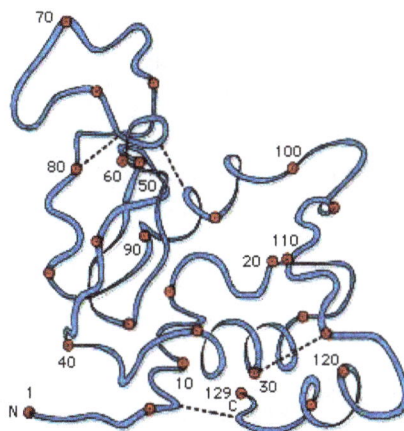

Lysozyme; protein conformation

The simplified structure of lysozyme from hen's egg white has a single peptide chain of 129 amino acids. The amino acid residues are numbered from the terminal α group (N) to the terminal carboxyl group (C). Circles indicate every fifth residue, and every tenth residue is numbered. Broken lines indicate the four disulfide bridges. Alpha-helices are visible in the ranges 25 to 35, 90 to 100, and 120 to 125.

Other Approaches to the Determination of Protein Structure

None of the several other physical methods that have been used to obtain information on the secondary and tertiary structure of proteins provides as much direct information as the X-ray diffraction technique. Most of the techniques, however, are much simpler than X-ray diffraction, which requires, for the resolution of the structure of one protein, many years of work and equipment such as electronic computers. Some of the simpler techniques are based on the optical properties of proteins—refractivity, absorption of light of different wavelengths, rotation of the plane polarized light at different wavelengths, and luminescence.

Spectrophotometric Behavior

Spectrophotometry of protein solutions (the measurement of the degree of absorbance of light by a protein within a specified wavelength) is useful within the range of visible light only with proteins that contain coloured prosthetic groups (the nonprotein components). Examples of such proteins include the red heme proteins of the blood, the purple pigments of the retina of the eye, green and yellow proteins that contain bile pigments, blue copper-containing proteins, and dark brown proteins called melanins. Peptide bonds, because of their carbonyl groups, absorb light energy at very short wavelengths (185–200 nanometres). The aromatic rings of phenylalanine, tyrosine, and tryptophan, however, absorb ultraviolet light between wavelengths of 280 and 290 nanometres. The absorbance of ultraviolet light by tryptophan is greatest, that of tyrosine is less, and that of phenylalanine is least. If the tyrosine or tryptophan content of the protein is known, therefore, the concentration of the protein solution can be determined by measuring its absorbance between 280 and 290 nanometres.

Optical Activity

It will be recalled that the amino acids, with the exception of glycine, exhibit optical activity. It is not surprising, therefore, that proteins also are optically active. They are usually levorotatory (i.e., they rotate the plane of polarization to the left) when polarized light of wavelengths in the visible range is used. Although the specific rotation (a function of the concentration of a protein solution and the distance the light travels in it) of most L-amino acids varies from –30° to +30°, the amino acid cystine has a specific rotation of approximately –300°. Although the optical rotation of a protein depends on all of the amino acids of which it is composed, the most important ones are cystine and the aromatic amino acids phenylalanine, tyrosine, and tryptophan. The contribution of the other amino acids to the optical activity of a protein is negligibly small.

Chemical Reactivity of Proteins

Information on the internal structure of proteins can be obtained with chemical methods that reveal whether certain groups are present on the surface of the protein molecule and thus able to react or whether they are buried inside the closely folded peptide chains and thus are unable to react. The chemical reagents used in such investigations must be mild ones that do not affect the structure of the protein.

The reactivity of tyrosine is of special interest. It has been found, for example, that only three of the six tyrosines found in the naturally occurring enzyme ribonuclease can be iodinated (i.e., reacted to accept an iodine atom). Enzyme-catalyzed breakdown of iodinated ribonuclease is used to identify the peptides in which the iodinated tyrosines are present. The three tyrosines that can be iodinated lie on the surface of ribonuclease; the others, assumed to be inaccessible, are said to be buried in the molecule. Tyrosine can also be identified by using other techniques—e.g., treatment with diazonium compounds or tetranitromethane. Because the compounds formed are coloured, they can easily be detected when the protein is broken down with enzymes.

Cysteine can be detected by coupling with compounds such as iodoacetic acid or iodoacetamide; the reaction results in the formation of carboxymethylcysteine or carbamidomethylcysteine, which can be detected by amino acid determination of the peptides containing them. The imidazole groups of certain histidines can also be located by coupling with the same reagents under different conditions. Unfortunately, few other amino acids can be labelled without changes in the secondary and tertiary structure of the protein.

Association of Protein Subunits

Many proteins with molecular weights of more than 50,000 occur in aqueous solutions as complexes: dimers, tetramers, and higher polymers—i.e., as chains of two, four, or more repeating basic structural units. The subunits, which are called monomers or protomers, usually are present as an even number. Less than 10 percent of the polymers have been found to have an odd number of monomers. The arrangement of the subunits is thought to be regular and may be cyclic, cubic, or tetrahedral. Some of the small proteins also contain subunits. Insulin, for example, with a molecular weight of about 6,000, consists of two peptide chains linked to each other by disulfide bridges ($-S-S-$). Similar interchain disulfide bonds have been found in the immunoglobulins. In other proteins, hydrogen bonds and hydrophobic bonds (resulting from the interaction between the amino acid side chains of valine, leucine, isoleucine, and phenylalanine) cause the formation of aggregates of the subunits. The subunits of some proteins are identical; those of others differ. Hemoglobin is a tetramer consisting of two α-chains and two β-chains.

Protein Denaturation

When a solution of a protein is boiled, the protein frequently becomes insoluble—i.e., it is denatured—and remains insoluble even when the solution is cooled. The denaturation of the proteins of egg white by heat—as when boiling an egg—is an example of irreversible denaturation. The denatured protein has the same primary structure as the original, or native, protein. The weak forces between charged groups and the weaker forces of mutual attraction of nonpolar groups are disrupted at elevated temperatures, however; as a result, the tertiary structure of the protein is lost. In some instances the original structure of the protein can be regenerated; the process is called renaturation.

Denaturation can be brought about in various ways. Proteins are denatured by treatment with alkaline or acid, oxidizing or reducing agents, and certain organic solvents. Interesting among denaturing agents are those that affect the secondary and tertiary structure without affecting the primary structure. The agents most frequently used for this purpose are urea and guanidinium chloride. These molecules, because of their high affinity for peptide bonds, break the hydrogen bonds and the salt bridges between positive and negative side chains, thereby abolishing the tertiary structure of the peptide chain. When denaturing agents are removed from a protein solution, the native protein re-forms in many cases. Denaturation can also be accomplished by reduction of the disulfide bonds of cystine—i.e., conversion of the disulfide bond ($-S-S-$) to two sulfhydryl groups ($-SH$). This, of course, results in the formation of two cysteines. Reoxidation of the cysteines by exposure to air sometimes regenerates the native protein. In other cases, however, the wrong cysteines become bound to each other, resulting in a different protein. Finally, denaturation can also be accomplished by exposing proteins to organic solvents such as ethanol or acetone. It is believed that the organic solvents interfere with the mutual attraction of nonpolar groups.

Some of the smaller proteins, however, are extremely stable, even against heat; for example, solutions of ribonuclease can be exposed for short periods of time to temperatures of 90 °C (194 °F) without undergoing significant denaturation. Denaturation does not involve identical changes in protein molecules. A common property of denatured proteins, however, is the loss of biological activity—e.g., the ability to act as enzymes or hormones.

Although denaturation had long been considered an all-or-none reaction, it is now thought that many intermediary states exist between native and denatured protein. In some instances, however, the breaking of a key bond could be followed by the complete breakdown of the conformation of the native protein.

Although many native proteins are resistant to the action of the enzyme trypsin, which breaks down proteins during digestion, they are hydrolyzed by the same enzyme after denaturation. The peptide bonds that can be split by trypsin are inaccessible in the native proteins but become accessible during denaturation. Similarly, denatured proteins give more intense colour reactions for tyrosine, histidine, and arginine than do the same proteins in the native state. The increased accessibility of reactive groups of denatured proteins is attributed to an unfolding of the peptide chains.

If denaturation can be brought about easily and if renaturation is difficult, how is the native conformation of globular proteins maintained in living organisms, in which they are produced stepwise, by incorporation of one amino acid at a time? Experiments on the biosynthesis of proteins from amino acids containing radioactive carbon or heavy hydrogen reveal that the protein molecule grows stepwise from the N terminus to the C terminus; in each step a single amino acid residue is incorporated. As soon as the growing peptide chain contains six or seven amino acid residues, the side chains interact with each other and thus cause deviations from the straight or β-chain configuration. Depending on the nature of the side chains, this may result in the formation of an α-helix or of loops closed by hydrogen bonds or disulfide bridges. The final conformation is probably frozen when the peptide chain attains a length of 50 or more amino acid residues.

Conformation of Proteins in Interfaces

Like many other substances with both hydrophilic and hydrophobic groups, soluble proteins tend to migrate into the interface between air and water or oil and water; the term oil here means a hydrophobic liquid such as benzene or xylene. Within the interface, proteins spread, forming thin films. Measurements of the surface tension, or interfacial tension, of such films indicate that tension is reduced by the protein film. Proteins, when forming an interfacial film, are present as a monomolecular layer—i.e., a layer one molecule in height. Although it was once thought that globular protein molecules unfold completely in the interface, it has now been established that many proteins can be recovered from films in the native state. The application of lateral pressure on a protein film causes it to increase in thickness and finally to form a layer with a height corresponding to the diameter of the native protein molecule. Protein molecules in an interface, because of Brownian motions (molecular vibrations), occupy much more space than do those in the film after the application of pressure. The Brownian motion of compressed molecules is limited to the two dimensions of the interface, since the protein molecules cannot move upward or downward.

The motion of protein molecules at the air–water interface has been used to determine the molecular weight of proteins. The technique involves measuring the force exerted by the protein layer on a barrier.

When a protein solution is vigorously shaken in air, it forms a foam, because the soluble proteins migrate into the air–water interface and persist there, preventing or slowing the reconversion of the foam into a homogeneous solution. Some of the unstable, easily modified proteins are denatured when spread in the air–water interface. The formation of a permanent foam when egg white is vigorously stirred is an example of irreversible denaturation by spreading in a surface.

Classification of Proteins

Classification by Solubility

proteins are essentially polypeptides consisting of many amino acids, an attempt was made to classify proteins according to their chemical and physical properties, because the biological function of proteins had not yet been established. (The protein character of enzymes was not proved until the 1920s.) Proteins were classified primarily according to their solubility in a number of solvents. This classification is no longer satisfactory, however, because proteins of quite different structure and function sometimes have similar solubilities; conversely, proteins of the same function and similar structure sometimes have different solubilities. The terms associated with the old classification, however, are still widely used.

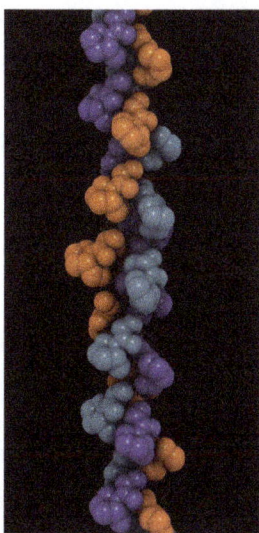

Collagen: Collagen molecule.

Albumins are proteins that are soluble in water and in water half-saturated with ammonium sulfate. On the other hand, globulins are salted out (i.e., precipitated) by half-saturation with ammonium sulfate. Globulins that are soluble in salt-free water are called pseudoglobulins; those insoluble in salt-free water are euglobulins. Both prolamins and glutelins, which are plant proteins, are insoluble in water; the prolamins dissolve in 50 to 80 percent ethanol, the glutelins in acidified or alkaline solution. The term protamine is used for a number of proteins in fish sperm that consist of approximately 80 percent arginine and therefore are strongly alkaline. Histones, which are less alkaline, apparently occur only in cell nuclei, where they are bound to nucleic acids. The term scleroproteins has been used for the insoluble proteins of animal organs. They include keratin, the insoluble protein of certain epithelial tissues such as the skin or hair, and collagen, the protein of the connective tissue. A large group of proteins has been called conjugated proteins, because they are complex molecules of protein consisting of protein and nonprotein moieties. The nonprotein portion is called the prosthetic group. Conjugated proteins can be subdivided into mucoproteins, which, in addition to protein, contain carbohydrate; lipoproteins, which contain lipids; phosphoproteins, which are rich in phosphate; chromoproteins, which contain pigments such as iron-porphyrins, carotenoids, bile pigments, and melanin; and finally, nucleoproteins, which contain nucleic acid.

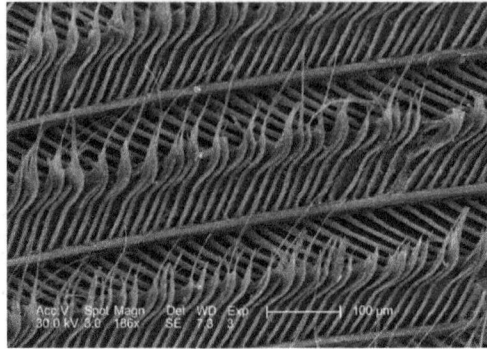

Keratin: Scanning electron micrograph showing strands
of keratin in a feather, magnified 186×.

The weakness of the above classification lies in the fact that many, if not all, globulins contain small amounts of carbohydrate; thus there is no sharp borderline between globulins and mucoproteins. Moreover, the phosphoproteins do not have a prosthetic group that can be isolated; they are merely proteins in which some of the hydroxyl groups of serine are phosphorylated (i.e., contain phosphate). Finally, the globulins include proteins with quite different roles—enzymes, antibodies, fibrous proteins, and contractile proteins.

Classification by Biological Functions

In view of the unsatisfactory state of the old classification, it is preferable to classify the proteins according to their biological function. Such a classification is far from ideal, however, because one protein can have more than one function. The contractile protein myosin, for example, also acts as an ATPase (adenosine triphosphatase), an enzyme that hydrolyzes adenosine triphosphate (removes a phosphate group from ATP by introducing a water molecule). Another problem with functional classification is that the definite function of a protein frequently is not known. A protein cannot be called an enzyme as long as its substrate (the specific compound upon which it acts) is not known. It cannot even be tested for its enzymatic action when its substrate is not known.

Special Structure and Function of Proteins

Despite its weaknesses, a functional classification is used here in order to demonstrate, whenever possible, the correlation between the structure and function of a protein. The structural, fibrous proteins are presented first, because their structure is simpler than that of the globular proteins and more clearly related to their function, which is the maintenance of either a rigid or a flexible structure.

Structural Proteins Scleroproteins

Collagen

Collagen is the structural protein of bones, tendons, ligaments, and skin. For many years collagen was considered to be insoluble in water. Part of the collagen of calf skin, however, can be extracted with citrate buffer at pH 3.7. A precursor of collagen called procollagen is converted in the body into collagen. Procollagen has a molecular weight of 120,000. Cleavage of one or a few peptide bonds of procollagen yields collagen, which has three subunits, each with a molecular weight of

95,000; therefore, the molecular weight of collagen is 285,000 (3 × 95,000). The three subunits are wound as spirals around an elongated straight axis. The length of each subunit is 2,900 angstroms, and its diameter is approximately 15 angstroms. The three chains are staggered, so that the trimer has no definite terminal limits.

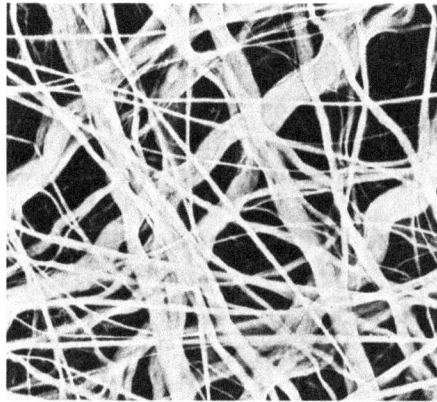

Randomly oriented collagenous fibres of varying size in a thin spread of loose areolar connective tissue (magnified about 370 ×).

Collagen differs from all other proteins in its high content of proline and hydroxyproline. Hydroxyproline does not occur in significant amounts in any other protein except elastin. Most of the proline in collagen is present in the sequence glycine–proline-X, in which X is frequently alanine or hydroxyproline. Collagen does not contain cystine or tryptophan and therefore cannot substitute for other proteins in the diet. The presence of proline causes kinks in the peptide chain and thus reduces the length of the amino acid unit from 3.7 angstroms in the extended chain of the β-structure to 2.86 angstroms in the collagen chain. In the intertwined triple helix, the glycines are inside, close to the axis; the prolines are outside.

Native collagen resists the action of trypsin but is hydrolyzed by the bacterial enzyme collagenase. When collagen is boiled with water, the triple helix is destroyed, and the subunits are partially hydrolyzed; the product is gelatin. The unfolded peptide chains of gelatin trap large amounts of water, resulting in a hydrated molecule.

When collagen is treated with tannic acid or with chromium salts, cross links form between the collagen fibres, and it becomes insoluble; the conversion of hide into leather is based on this tanning process. The tanned material is insoluble in hot water and cannot be converted to gelatin. On exposure to water at 62° to 63° C (144° to 145° F), however, the cross links formed by the tanning agents collapse, and the leather contracts irreversibly to about one-third its original volume.

Collagen seems to undergo an aging process in living organisms that may be caused by the formation of cross links between collagen fibres. They are formed by the conversion of some lysine side chains to aldehydes (compounds with the general structure RCHO), and the combination of the aldehydes with the ε-amino groups of intact lysine side chains. The protein elastin, which occurs in the elastic fibres of connective tissue, contains similar cross links and may result from the combination of collagen fibres with other proteins. When cross-linked collagen or elastin is degraded, products of the cross-linked lysine fragments, called desmosins and isodesmosins, are formed.

Keratin

Keratin, the structural protein of epithelial cells in the outermost layers of the skin, has been isolated from hair, nails, hoofs, and feathers. Keratin is completely insoluble in cold or hot water; it is not attacked by proteolytic enzymes (i.e., enzymes that break apart, or lyse, protein molecules), and therefore cannot replace proteins in the diet. The great stability of keratin results from the numerous disulfide bonds of cystine. The amino acid composition of keratin differs from that of collagen. Cystine may account for 24 percent of the total amino acids. The peptide chains of keratin are arranged in approximately equal amounts of antiparallel and parallel pleated sheets, in which the peptide chains are linked to each other by hydrogen bonds between the carbonyl and imino groups.

Reduction of the disulfide bonds to sulfhydryl groups results in dissociation of the peptide chains, the molecular weight of which is 25,000 to 28,000 each. The formation of permanent waves in the beauty treatment of hair is based on partial reduction of the disulfide bonds of hair keratin by thioglycol, or some other mild reducing agent, and subsequent oxidation of the sulfhydryl groups (−SH) in the reoriented hair to disulfide bonds (−S−S−) by exposure to the oxygen of the air.

The length of keratin fibres depends on their water content. They can bind approximately 16 percent of water; this hydration is accompanied by an increase in the length of the fibres of 10 to 12 percent.

The most thoroughly investigated keratin is hair keratin, particularly that of wool. It consists of a mixture of peptides with high and low cystine content. When wool is heated in water to about 90° C (190° F), it shrinks irreversibly. This is attributed to the breakage of hydrogen bonds and other noncovalent bonds; disulfide bonds do not seem to be affected.

Others

The most thoroughly investigated scleroprotein has been fibroin, the insoluble material of silk. The raw silk comprising the cocoon of the silkworm consists of two proteins. One, sericin, is soluble in hot water; the other, fibroin, is not. The amino acid composition of the latter differs from that of all other proteins. It contains large amounts of glycine, alanine, tyrosine, and serine; small amounts of the other amino acids; and no sulfur-containing ones. The peptide chains are arranged in antiparallel β-structures. Fibroin is partly soluble in concentrated solutions of lithium thiocyanate or in mixtures of cupric salts and ethylene diamine. Such solutions contain a protein of molecular weight 170,000, which is a dimer of two subunits.

Little is known about either the scleroproteins of the marine sponges or the insoluble proteins of the cellular membranes of animal cells. Some of the membranes are soluble in detergents; others, however, are detergent-insoluble.

Muscle Proteins

The total amount of muscle proteins in mammals, including humans, exceeds that of any other protein. About 40 percent of the body weight of a healthy human adult weighing about 70 kilograms (150 pounds) is muscle, which is composed of about 20 percent muscle protein. Thus, the human body contains about 5 to 6 kilograms (11 to 13 pounds) of muscle protein. An albumin-like fraction of these proteins, originally called myogen, contains various enzymes—phosphorylase, aldolase, glyceraldehyde phosphate dehydrogenase, and others; it does not seem to be involved in contraction. The globulin fraction contains myosin, the contractile protein, which also occurs in blood platelets, small bodies found in blood. Similar contractile substances occur in other contractile structures; for example, in the cilia or flagella (whiplike organs of locomotion) of bacteria and protozoans. In contrast to the scleroproteins, the contractile proteins are soluble in salt solutions and susceptible to enzymatic digestion.

The energy required for muscle contraction is provided by the oxidation of carbohydrates or lipids. The term mechanochemical reaction has been used for this conversion of chemical into mechanical energy. The molecular process underlying the reaction is known to involve the fibrous muscle proteins, the peptide chains of which undergo a change in conformation during contraction.

Myosin, which can be removed from fresh muscle by adding it to a chilled solution of dilute potassium chloride and sodium bicarbonate, is insoluble in water. Myosin, solutions of which are highly viscous, consists of an elongated—probably double-stranded—peptide chain, which is coiled at both ends in such a way that a terminal globule is formed. The length of the molecule is approximately 160 nanometres and its average diameter 2.6 nanometres. The equivalent weight of each of the two terminal globules is approximately 30,000; the molecular weight of myosin is close to 500,000. Trypsin splits myosin into large fragments called meromyosin. Myosin contains many amino acids with positively and negatively charged side chains; they form 18 and 16 percent, respectively, of the total number of amino acids. Myosin catalyzes the hydrolytic cleavage of ATP (adenosine triphosphate). A smaller protein with properties similar to those of myosin is tropomyosin. It has a molecular weight of 70,000 and dimensions of 45 by 2 nanometres. More than 90 percent of its peptide chains are present in the α-helix form.

Myosin combines easily with another muscle protein called actin, the molecular weight of which is about 50,000; it forms 12 to 15 percent of the muscle proteins. Actin can exist in two forms—one, G-actin, is globular; the other, F-actin, is fibrous. Actomyosin is a complex molecule formed by one molecule of myosin and one or two molecules of actin. In muscle, actin and myosin filaments are oriented parallel to each other and to the long axis of the muscle. The actin filaments are linked to each other lengthwise by fine threads called S filaments. During contraction the S filaments shorten, so that the actin filaments slide toward each other, past the myosin filaments, thus causing a shortening of the muscle.

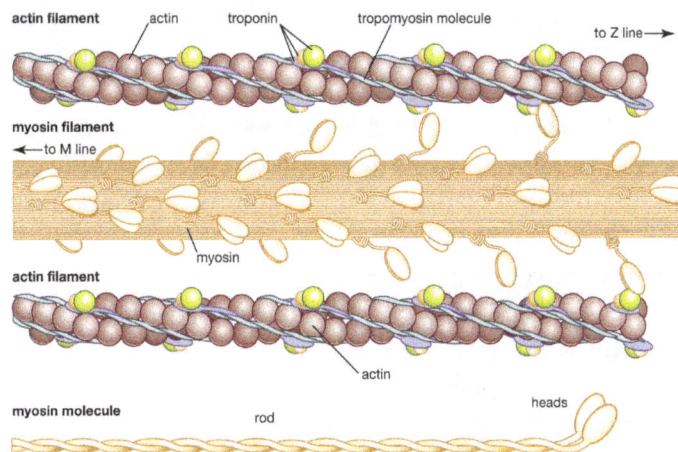

Muscle: actin and myosin
The structure of actin and myosin filaments.

Fibrinogen and Fibrin

Fibrinogen, the protein of the blood plasma, is converted into the insoluble protein fibrin during the clotting process. The fibrinogen-free fluid obtained after removal of the clot, called blood serum, is blood plasma minus fibrinogen. The fibrinogen content of the blood plasma is 0.2 to 0.4 percent.

Fibrinogen can be precipitated from the blood plasma by half-saturation with sodium chloride. Fibrinogen solutions are highly viscous and show strong flow birefringence. In electron micrographs the molecules appear as rods with a length of 47.5 nanometres and a diameter of 1.5 nanometres; in addition, two terminal and a central nodule are visible. The molecular weight is 340,000. An unusually high percentage, about 36 percent, of the amino acid side chains are positively or negatively charged.

The clotting process is initiated by the enzyme thrombin, which catalyzes the breakage of a few peptide bonds of fibrinogen; as a result, two small fibrinopeptides with molecular weights of 1,900 and 2,400 are released. The remainder of the fibrinogen molecule, a monomer, is soluble and stable at pH values less than 6 (i.e., in acid solutions). In neutral solution (pH 7) the monomer is converted into a larger molecule, insoluble fibrin; this results from the formation of new peptide bonds. The newly formed peptide bonds form intermolecular and intramolecular cross links, thus giving rise to a large clot, in which all molecules are linked to each other. Clotting, which takes place only in the presence of calcium ions, can be prevented by compounds such as oxalate or citrate, which have a high affinity for calcium ions.

Albumins, Globulins and other Soluble Proteins

The blood plasma, the lymph, and other animal fluids usually contain one to seven grams of protein per 100 millilitres of fluid, which includes small amounts of hundreds of enzymes and a large number of protein hormones.

Proteins of Blood Serum

Human blood serum contains about 7 percent protein, two-thirds of which is in the albumin fraction; the other third is in the globulin fraction. Electrophoresis of serum reveals a large albumin peak and three smaller globulin peaks, the alpha-, beta-, and gamma-globulins. The amounts of alpha-, beta-, and gamma-globulin in normal human serum are approximately 1.5, 1.9, and 1.1 percent, respectively. Each globulin fraction is a mixture of many different proteins, as has been demonstrated by immunoelectrophoresis. In this method, serum from an animal (e.g., a rabbit) injected with human serum is allowed to diffuse into the four protein bands—albumin, alpha-, beta-, and gamma-globulin—obtained from the electrophoresis of human serum. Because the animal has previously been injected with human serum, its blood contains antibodies (substances formed in response to a foreign substance introduced into the body) against each of the human serum proteins; each antibody combines with the serum protein (antigen) that caused its formation in the animal. The result is the formation of about 20 regions of insoluble antigen-antibody precipitate, which appear as white arcs in the transparent gel of the electrophoresis medium. Each region corresponds to a different human serum protein.

Serum albumin is much less heterogeneous (i.e., contains fewer distinct proteins) than are the globulins; in fact, it is one of the few serum proteins that can be obtained in a crystalline form.

Serum albumin combines easily with many acidic dyes (e.g., Congo red and methyl orange); with bilirubin, the yellow bile pigment; and with fatty acids. It seems to act, in living organisms, as a carrier for certain biological substances. Present in blood serum in relatively high concentration, serum albumin also acts as a protective colloid, a protein that stabilizes other proteins. Albumin (molecular weight of 68,000) has a single free sulfhydryl ($-SH$) group, which on oxidation forms a disulfide bond with the sulfhydryl group of another serum albumin molecule, thus forming a dimer. The isoelectric point of serum albumin is pH 4.7.

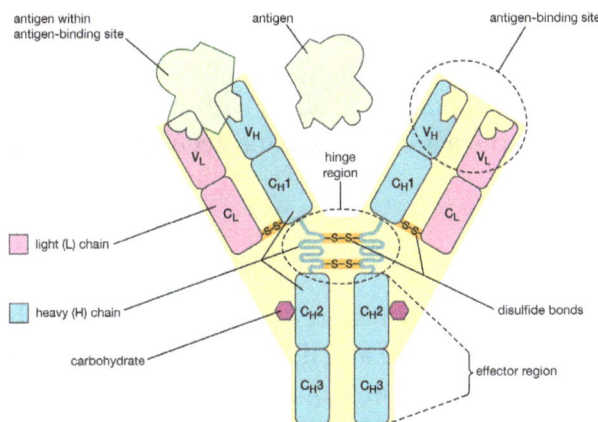

The four-chain structure of an antibody, or immunoglobulin, molecule

The basic unit is composed of two identical light (L) chains and two identical heavy (H) chains, which are held together by disulfide bonds to form a flexible Y shape. Each chain is composed of a variable (V) region and a constant (C) region.

The alpha-globulin fraction of blood serum is a mixture of several conjugated proteins. The best known are an α-lipoprotein (combination of lipid and protein) and two mucoproteins (combinations of carbohydrate and protein). One mucoprotein is called orosomucoid, or α1-acid glycoprotein; the other is called haptoglobin because it combines specifically with globin, the protein component of hemoglobin. Haptoglobin contains about 20 percent carbohydrate. The beta-globulin fraction of serum contains, in addition to lipoproteins and mucoproteins, two metal-binding proteins, transferrin and ceruloplasmin, which bind iron and copper, respectively. They are the principal iron and copper carriers of the blood.

The gamma-globulins are the most heterogeneous globulins. Although most have a molecular weight of approximately 150,000, that of some, called macroglobulins, is as high as 800,000. Because typical antibodies are of the same size and exhibit the same electrophoretic behaviour as γ-globulins, they are called immunoglobulins. The designation IgM or gamma M (γM) is used for the macroglobulins; the designation IgG or gamma G (γG) is used for γ–globulins of molecular weight 150,000.

Milk Proteins

Milk contains the following: an albumin, α-lactalbumin; a globulin, beta-lactoglobulin; and a phosphoprotein, casein. If acid is added to milk, casein precipitates. The remaining watery liquid (the supernatant solution), or whey, contains α-lactalbumin and β-lactoglobulin. Both have been obtained in crystalline form; in bovine milk, their molecular weights are approximately 14,000 and

18,400, respectively. Lactoglobulin also occurs as a dimer of molecular weight 37,000. Genetic variations can produce small variations in the amino acid composition of lactoglobulin. The amino acid composition and the tertiary structure of lactalbumin resemble that of lysozyme, an egg protein.

Casein is precipitated not only by the addition of acid but also by the action of the enzyme rennin, which is found in gastric juice. Rennin from calf stomachs is used to precipitate casein, from which cheese is made. Milk fat precipitates with casein; milk sugar, however, remains in the supernatant (whey). Casein is a mixture of several similar phosphoproteins, called α-, β-, γ-, and κ-casein, all of which contain some serine side chains combined with phosphoric acid. Approximately 75 percent of casein is α-casein. Cystine has been found only in κ-casein. In milk, casein seems to form polymeric globules (micelles) with radially arranged monomers, each with a molecular weight of 24,000; the acidic side chains occur predominantly on the surface of the micelle, rather than inside.

Egg Proteins

About 50 percent of the proteins of egg white are composed of ovalbumin, which is easily obtained in crystals. Its molecular weight is 46,000 and its amino acid composition differs from that of serum albumin. Other proteins of egg white are conalbumin, lysozyme, ovoglobulin, ovomucoid, and avidin. Lysozyme is an enzyme that hydrolyzes the carbohydrates found in the capsules certain bacteria secrete around themselves; it causes lysis (disintegration) of the bacteria. The molecular weight of lysozyme is 14,100. Its three-dimensional structure is similar to that of α-lactalbumin, which stimulates the formation of lactose by the enzyme lactose synthetase. Lysozyme has also been found in the urine of patients suffering from leukemia, meningitis, and renal disease.

Avidin is a glycoprotein that combines specifically with biotin, a vitamin. In animals fed large amounts of raw egg white, the action of avidin results in "egg-white injury." The molecular weight of avidin, which forms a tetramer, is 16,200. Its amino acid sequence is known.

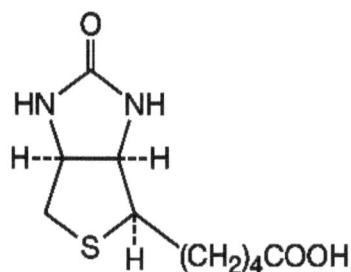

Egg-yolk proteins contain a mixture of lipoproteins and livetins. The latter are similar to serum albumin, α-globulin, and β-globulin. The yolk also contains a phosphoprotein, phosvitin. Phosvitin, which has also been found in fish sperm, has a molecular weight of 40,000 and an unusual amino acid composition; one third of its amino acids are phosphoserine.

Protamines and Histones

Protamines are found in the sperm cells of fish. The most thoroughly investigated protamines are salmine from salmon sperm and clupeine from herring sperm. The protamines are bound to deoxyribonucleic acid (DNA), forming nucleoprotamines. The amino acid composition of the protamines is simple; they contain, in addition to large amounts of arginine, small amounts of five or six other amino acids. The composition of the salmine molecule, for example, is: Arg51, Ala4, Val4, Ile1, Pro7, and Ser6, in which the subscript numbers indicate the number of each amino acid in the molecule. Because of the high arginine content, the isoelectric points of the protamines are at pH values of 11 to 12; i.e., the protamines are alkaline. The molecular weights of salmine and clupeine are close to 6,000. All of the protamines investigated thus far are mixtures of several similar proteins.

The histones are less basic than the protamines. They contain high amounts of either lysine or arginine and small amounts of aspartic acid and glutamic acid. Histones occur in combination with DNA as nucleohistones in the nuclei of the body cells of animals and plants, but not in animal sperm. The molecular weights of histones vary from 10,000 to 22,000. In contrast to the protamines, the histones contain most of the 20 amino acids, with the exception of tryptophan and the sulfur-containing ones. Like the protamines, histone preparations are heterogeneous mixtures. The amino acid sequence of some of the histones has been determined.

Plant Proteins

Plant proteins, mostly globulins, have been obtained chiefly from the protein-rich seeds of cereals and legumes. Small amounts of albumins are found in seeds. The best known globulins, insoluble in water, can be extracted from seeds by treatment with 2 to 10 percent solutions of sodium chloride. Many plant globulins have been obtained in crystalline form; they include edestin from hemp, molecular weight 310,000; amandin from almonds, 330,000; concanavalin A (42,000) and B (96,000); and canavalin (113,000) from jack beans. They are polymers of smaller subunits; edestin, for example, is a hexamer of a subunit with a molecular weight of 50,000, and concanavalin B a trimer of a subunit with a molecular weight of 30,000. After extraction of lipids from cereal seeds by ether and alcohol, further extraction with water containing 50 to 80 percent of alcohol yields proteins that are insoluble in water but soluble in water–ethanol mixtures and have been called prolamins. Their solubility in aqueous ethanol may result from their high proline and glutamine content. Gliadin, the prolamin from wheat, contains 14 grams of proline and 46 grams of glutamic acid in 100 grams of protein; most of the glutamic acid is in the form of glutamine. The total amounts of the basic amino acids (arginine, lysine, and histidine) in gliadin are only 5 percent of the weight of gliadin. Because the glysine content is either low or nonexistent, human populations dependent on grain as a sole protein source suffer from lysine deficiency.

Conjugated Proteins

Combination of Proteins with Prosthetic Groups

The link between a protein molecule and its prosthetic group is a covalent bond (an electron-sharing bond) in the glycoproteins, the biliproteins, and some of the heme proteins. In lipoproteins, nucleoproteins, and some heme proteins, the two components are linked by noncovalent bonds; the bonding results from the same forces that are responsible for the tertiary structure of proteins: hydrogen bonds, salt bridges between positively and negatively charged groups, disulfide bonds, and mutual interaction of hydrophobic groups. In the metalloproteins (proteins with a metal element as a prosthetic group), the metal ion usually forms a centre to which various groups are bound.

Some of the conjugated proteins have been mentioned in preceding sections because they occur in the blood serum, in milk, and in eggs.

Mucoproteins and Glycoproteins

The prosthetic groups in mucoproteins and glycoproteins are oligosaccharides (carbohydrates consisting of a small number of simple sugar molecules) usually containing from four to 12 sugar molecules; the most common sugars are galactose, mannose, glucosamine, and galactosamine. Xylose, fucose, glucuronic acid, sialic acid, and other simple sugars sometimes also occur. Some mucoproteins contain 20 percent or more of carbohydrate, usually in several oligosaccharides attached to different parts of the peptide chain. The designation mucoprotein is used for proteins with more than 3 to 4 percent carbohydrate; if the carbohydrate content is less than 3 percent, the protein is sometimes called a glycoprotein or simply a protein.

Mucoproteins, highly viscous proteins originally called mucins, are found in saliva, in gastric juice, and in other animal secretions. Mucoproteins occur in large amounts in cartilage, synovial fluid (the lubricating fluid of joints and tendons), and egg white. The mucoprotein of cartilage is formed by the combination of collagen with chondroitinsulfuric acid, which is a polymer of either glucuronic or iduronic acid and acetylhexosamine or acetylgalactosamine. It is not yet clear whether or not chondroitinsulfate is bound to collagen by covalent bonds.

Lipoproteins and Proteolipids

The bond between the protein and the lipid portion of lipoproteins and proteolipids is a noncovalent one. It is thought that some of the lipid is enclosed in a meshlike arrangement of peptide chains and becomes accessible for reaction only after the unfolding of the chains by denaturing agents. Although lipoproteins in the α- and β-globulin fraction of blood serum are soluble in water (but insoluble in organic solvents), some of the brain lipoproteins, because they have a high lipid content, are soluble in organic solvents; they are called proteolipids. The β-lipoprotein of human blood serum is a macroglobulin with a molecular weight of about 1,300,000, 70 percent of which is lipid; of the lipid, about 30 percent is phospholipid and 40 percent cholesterol and compounds derived from it. Because of their lipid content, the lipoproteins have the lowest density (mass per unit volume) of all proteins and are usually classified as low- and high-density lipoproteins (LDL and HDL).

Coloured lipoproteins are formed by the combination of protein with carotenoids. Crustacyanin, the pigment of lobsters, crayfish, and other crustaceans, contains astaxanthin, which is a compound derived from carotene. Among the most interesting of the coloured lipoproteins are the pigments of the retina of the eye. They contain retinal, which is a compound derived from carotene and which is formed by the oxidation of vitamin A. In rhodopsin, the red pigment of the retina, the aldehyde group ($-CHO$) of retinal forms a covalent bond with an amino ($-NH_2$) group of opsin, the protein carrier. Colour vision is mediated by the presence of several visual pigments in the retina that differ from rhodopsin either in the structure of retinal or in that of the protein carrier.

Metalloproteins

Proteins in which heavy metal ions are bound directly to some of the side chains of histidine, cysteine, or some other amino acid are called metalloproteins. Two metalloproteins, transferri-nand ceruloplasmin, occur in the globulin fractions of blood serum; they act as carriers of iron and copper, respectively. Transferrin has a molecular weight of about 80,000 and consists of two identical subunits, each of which contains one ferric ion (Fe^{3+}) that seems to be bound to tyrosine. Several genetic variants of transferrin are known to occur in humans. Another iron protein, fer-ritin, which contains 20 to 22 percent iron, is the form in which iron is stored in animals; it has been obtained in crystalline form from liver and spleen. A molecule consisting of 20 subunits, its molecular weight is approximately 480,000. The iron can be removed by reduction from the fer-ric (Fe^{3+}) to the ferrous (Fe^{2+}) state. The iron-free protein, apoferritin, is synthesized in the body before the iron is incorporated.

Green plants and some photosynthetic and nitrogen-fixing bacteria (i.e., bacteria that convert atmospheric nitrogen, N_2, into amino acids and proteins) contain various ferredoxins. They are small proteins containing 50 to 100 amino acids and a chain of iron and disulfide units (FeS_2), in which some of the sulfur atoms are contributed by cysteine; others are sulfide ions (S^{2-}). The number of FeS_2 units per ferredoxin molecule varies from five in the ferredoxin of spinach to 10 in the ferredoxin of certain bacteria. Ferredoxins act as electron carriers in photosynthesis and in nitrogen fixation.

Ceruloplasmin is a copper-containing globulin that has a molecular weight of 151,000; the molecule consists of eight subunits, each containing one copper ion. Ceruloplasmin is the principal carrier of copper in organisms, although copper can also be transported by the iron-containing globu-lin transferrin. Another copper-containing protein, copper-zinc superoxide dismutase (formerly known as erythrocuprein), has been isolated from red blood cells; it has also been found in the liver and in the brain. The molecule, which consists of two subunits of similar size, contains cop-per ions and zinc ions. Because of their copper content, ceruloplasmin and copper-zinc superoxide dismutase possess catalytic activity in oxidation-reduction reactions.

Many animal enzymes contain zinc ions, which are usually bound to the sulfur of cysteine. Horse kidneys contain the protein metallothionein, which contain zinc and cadmium; both are bound to sulfur. A vanadium-protein complex (hemovanadin) has been found in surprisingly high amounts in yellowish-green cells (vanadocytes) of tunicates, which are marine invertebrates.

Heme Proteins and other Chromoproteins

Although the heme proteins contain iron, they are usually not classified as metalloproteins, be-cause their prosthetic group is an iron-porphyrin complex in which the iron is bound very firmly. The intense red or brown colour of the heme proteins is not caused by iron but by porphyrin, a complex cyclic structure. All porphyrin compounds absorb light intensely at or close to 410 nano-metres. Porphyrin consists of four pyrrole rings (five-membered closed structures containing one nitrogen and four carbon atoms) linked to each other by methine groups ($-CH=$). The iron atom is kept in the centre of the porphyrin ring by interaction with the four nitrogen atoms. The iron atom can combine with two other substituents; in oxyhemoglobin, one substituent is a histidine of the protein carrier, the other is an oxygenmolecule. In some heme proteins, the protein is also bound covalently to the side chains of porphyrin.

The chromoprotein melanin, a pigment found in dark skin, dark hair, and melanotic tumours, occurs in every major group of living organisms and appears to be remarkably diverse in structure. In humans, melanin produced by melanocytes may be dark brown (eumelanin) or pale red or yellowish (phaeomelanin). The different types are synthesized via different pathways, though they share the same initial step—the oxidation of tyrosine.

Green chromoproteins called biliproteins are found in many insects, such as grasshoppers, and also in the eggshells of many birds. The biliproteins are derived from the bile pigment biliverdin, which in turn is formed from porphyrin; biliverdin contains four pyrrole rings and three of the four methine groups of porphyrin. Large amounts of biliproteins have been found in red algae and blue-green algae; the red protein is called phycoerythrin, the blue one phycocyanobilin.

Blue-green algae in Morning Glory Pool, Yellowstone National Park, Wyoming.

Nucleoproteins

When a protein solution is mixed with a solution of a nucleic acid, the phosphoric acidcomponent of the nucleic acid combines with the positively charged ammonium groups ($-NH_3^+$) of the protein to form a protein–nucleic acid complex. The nucleus of a cell contains predominantly deoxyribonucleic acid (DNA) and the cytoplasm predominantly ribonucleic acid (RNA); both parts of the cell also contain protein. Protein–nucleic acid complexes, therefore, form in living cells.

The only nucleoproteins for which some evidence for specificity exists are nucleoprotamines, nucleohistones, and some RNA and DNA viruses. The nucleoprotamines are the form in which protamines occur in the sperm cells of fish; the histones of the thymus and of pea seedlings and other plant material apparently occur predominantly as nucleohistones. Both nucleoprotamines and nucleohistones contain only DNA.

Some of the simplest viruses consist of a specific RNA, which is coated by protein. One of the best known RNA viruses, tobacco mosaic virus (TMV), has the shape of a rod. RNA comprisesonly 5.1 percent of the mass of the virus. The complete sequence of the virus protein, which consists of about 2,130 identical peptide chains, each containing 158 amino acids, has been determined. The protein is arranged in a spiral around the RNA core.

DNA has been found in most bacterial viruses (bacteriophages) and in some animal viruses. As in TMV, the core of DNA is surrounded by protein. Phage protein is a mixture of enzymes and therefore cannot be considered as the protein portion of only one nucleoprotein.

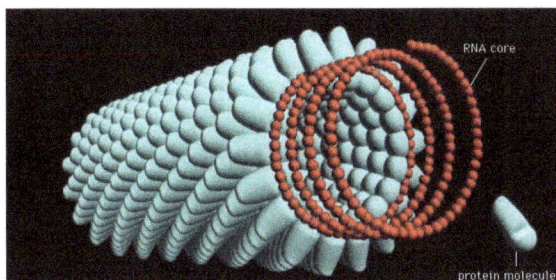

Schematic structure of the tobacco mosaic virus. The cutaway section shows the helical ribonucleic acid a ssociated with protein molecules in a ratio of three nucleotides per protein molecule.

Respiratory Proteins

Hemoglobin

Hemoglobin is the oxygen carrier in all vertebrates and some invertebrates. In oxyhemoglobin (HbO_2), which is bright red, the ferrous ion (Fe^{2+}) is bound to the four nitrogen atoms of porphyrin; the other two substituents are an oxygen molecule and the histidine of globin, the protein component of hemoglobin. Deoxyhemoglobin (deoxy-Hb), as its name implies, is oxyhemoglobin minus oxygen (i.e., reduced hemoglobin); it is purple in colour. Oxidation of the ferrous ion of hemoglobin yields a ferric compound, methemoglobin, sometimes called hemiglobin or ferrihemoglobin. The oxygen of oxyhemoglobin can be displaced by carbon monoxide, for which hemoglobin has a much greater affinity, preventing oxygen from reaching the body tissues.

The hemoglobins of all mammals, birds, and many other vertebrates are tetramers of two α- and two β-chains. The molecular weight of the tetramer is 64,500; the molecular weight of the α- and β-chains is approximately 16,100 each, and the four subunits are linked to each other by noncovalent interactions. If hemin (the ferric porphyrin component) is removed from globin (the protein component), two molecules of globin, each consisting of one α- and one β-chain, are obtained; the molecular weight of globin is 32,200. In contrast to hemoglobin, globin is an unstable protein that is easily denatured. If native globin is incubated with a solution of hemin at pH values of 8 to 9, native hemoglobin is reconstituted. Myoglobin, the red pigment of mammalian muscles, is a monomer with a molecular weight of 16,000.

The mammalian hemoglobins differ from each other in their amino acid composition and therefore in their secondary and tertiary structure. Rat and horse hemoglobins crystallize very easily, but those of humans, cattle, and sheep, because they are more soluble, are difficult to crystallize. The shape of hemoglobin crystals varies in different species; moreover, decomposition and denaturation occur at different rates in different species. It was also found that the blood of human newborns contains two different hemoglobins: about 20 percent of their hemoglobin is an adult hemoglobin (hemoglobin A) and 80 percent is a fetal hemoglobin (hemoglobin F). Hemoglobin F persists in the infant for the first seven months of life. The same hemoglobin F has also been found in the blood of patients suffering from thalassemia, an anemia with a high incidence in regions surrounding the Mediterranean Sea. Hemoglobin F contains, as does hemoglobin A, two α-chains; the two β-chains, however, have been replaced by two quite different γ-chains.

The hemoglobins of some of the lowest fishes are monomers containing one iron atom per molecule. Hemoglobin-like respiratory proteins have been found in some invertebrates. The red hemoglobin of insects, mollusks, and protozoans is called erythrocruorin. It differs from vertebrate hemoglobin by its high molecular weight.

Although green plants contain no hemoglobin, a red protein, called leghemoglobin, has been discovered in the root nodules of leguminous plants. It seems to be produced by the nitrogen-fixing bacteria of the root nodules and may be involved in the reduction of atmospheric nitrogen to ammonia and amino acids.

Other Respiratory Proteins

A green respiratory protein, chlorocruorin, has been found in the blood of marine worms in the genera Serpula and Spirographis. It has the same high molecular weight as erythrocruorin but differs from hemoglobin in its prosthetic group. A red metalloprotein, hemerythrin, acts as a respiratory protein in marine worms of the phylum Sipuncula. The molecule consists of eight subunits with a molecular weight of 13,500 each. Hemerythrin contains no porphyrins and therefore is not a heme protein.

A metalloprotein containing copper is the respiratory protein of crustaceans (shrimps, crabs, etc.) and of some gastropods (snails). The protein, called hemocyanin, is pale yellow when not combined with oxygen, and blue when combined with oxygen. The molecular weights of hemocyanins vary from 300,000 to 9,000,000. Each animal investigated thus far apparently has a species-specific hemocyanin.

Protein Hormones

Some hormones that are products of endocrine glands are proteins or peptides, others are steroids. None of the hormones has any enzymatic activity. Each has a target organ in which it elicits some biological action—e.g., secretion of gastric or pancreatic juice, production of milk, production of steroid hormones. The mechanism by which the hormones exert their effects is not fully understood. Cyclic adenosine monophosphate is involved in the transmittance of the hormonal stimulus to the cells whose activity is specifically increased by the hormone.

Hormones of Thyroid Gland

Thyroglobulin, the active groups of which are two molecules of the iodine-containing compound thyroxine, has a molecular weight of 670,000. Thyroglobulin also contains thyroxine with two and three iodine atoms instead of four and tyrosine with one and two iodine atoms. Injection of the hormone causes an increase in metabolism; lack of it results in a slowdown.

Another hormone, calcitonin, which lowers the calcium level of the blood, occurs in the thyroid gland. The amino acid sequences of calcitonin from pig, beef, and salmon differ from human calcitonin in some amino acids. All of them, however, have the half-cystines (C) and the prolinamide (P) in the same position.

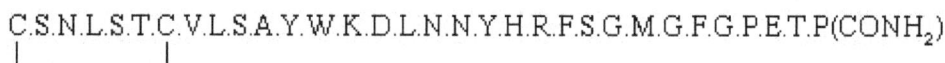

C.S.N.L.S.T.C.V.L.S.A.Y.W.K.D.L.N.N.Y.H.R.F.S.G.M.G.F.G.P.E.T.P(CONH$_2$)

Parathyroid hormone (parathormone), produced in small glands that are embedded in or lie behind the thyroid gland, is essential for maintaining the calcium level of the blood. A decrease in its production results in hypocalcemia (a reduction of calcium levels in the bloodstream below the normal range). Bovine parathormone has a molecular weight of 8,500; it contains no cystine or cysteine and is rich in aspartic acid, glutamic acid, or their amides.

Hormones of the Pancreas

Although the amino acid structure of insulin has been known since 1949, repeated attempts to synthesize it gave very poor yields because of the failure of the two peptide chains to combine forming the correct disulfide bridge. The ease of the biosynthesis of insulin is explained by the discovery in the pancreas of proinsulin, from which insulin is formed. The single peptide chain of proinsulin loses a peptide consisting of 33 amino acids and called the connecting peptide, or C peptide, during its conversion to insulin. The disulfide bridges of proinsulin connect the A and B chains.

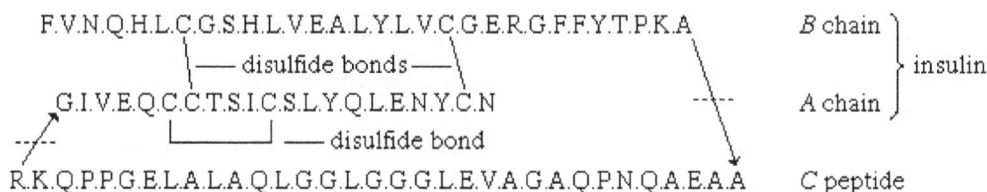

In aqueous solutions, insulin exists predominantly as a complex of six subunits, each of which contains an A and a B chain. The insulins of several species have been isolated and analyzed; their amino acid sequences have been found to differ somewhat, but all apparently contain the same disulfide bridges between the two chains.

Although the injection of insulin lowers the blood sugar, administration of glucagon, another pancreas hormone, raises the blood sugar level. Glucagon consists of a straight peptide chain of 29 amino acids. It has been synthesized; the synthetic product has the full biological activity of natural glucagon. The structure of glucagon is free of cystine and isoleucine.

The pituitary gland has an anterior lobe, a posterior lobe, and an intermediate portion; they differ in cellular structure and in the structure and action of the hormones they form. The posterior lobe produces two similar hormones, oxytocin and vasopressin. The former causes contraction of the pregnant uterus; the latter raises the blood pressure. Both are octapeptides formed by a ring of five amino acids (the two cystine halves count as one amino acid) and a side chain of three amino acids. The two cystine halves are linked to each other by a disulfide bond, and the C terminal amino acid is glycinamide. The structure has been established and confirmed. Human vasopressin differs from oxytocin in that isoleucine is replaced by phenylalanine and leucine by arginine.

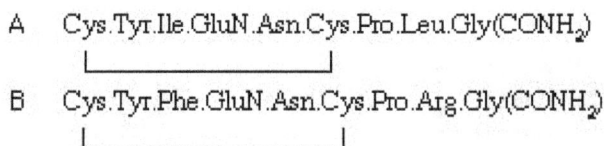

The intermediate part of the pituitary gland produces the melanocyte-stimulating hormone (MSH), which causes expansion of the pigmented melanophores (cells) in the skin of frogs and other batrachians. Two hormones, called α-MSH and β-MSH, have been prepared from hog pituitary glands. The first, α-MSH, consists of 13 amino acids; its N terminal serine is acetylated (i.e., the acetyl group, CH_3CO, of acetic acid is attached), and its C terminal valine residue is present as valinamide. The second, β-MSH, contains in its 18 amino acids many of those occurring in α-MSH.

$(CH_3CO)S.Y.S.M.E.H.F.R.W.G.K.P.V.(CONH_2)$ porcine α-MSH, melanocyte-stimulating hormone

D.S.G.P.Y.K.M.E.H.F.R.W.G.S.P.P.K.D   porcine β-MSH

A.E.K.K.D.E.G.P.Y.K.M.E.H.F.R.W.G.S.P.P.K.D   human β-MSH

S.Y.S.M.E.H.F.R.W.G.K.P.V.G.K.K.R.R.P.V.K.V.Y.P.D.G.A.E.D.Q.L.A.E.A.F.P.L.E.F   porcine β-corticotropin

The anterior pituitary lobe produces several protein hormones—a thyroid-stimulating hormone (thyrotropin), molecular weight 28,000; a lactogenic hormone, molecular weight 22,500; a growth hormone, molecular weight 21,500; a luteinizing hormone, molecular weight 30,000; and a follicle-stimulating hormone, molecular weight 29,000. The thyroid-stimulating hormone consists of α and β subunits with a composition similar to the subunits of luteinizing hormone. When separated, neither of the two subunits has hormonal activity; when combined, however, they regain about 50 percent of the original activity. The lactogenic hormone (prolactin) from sheep pituitary glands contains 190 amino acids. Their sequence has been elucidated; a similar peptide chain of 188 amino acids that has been synthesized not only has 10 percent of the biological activity of the natural hormone but also some activity of the growth hormone. The amino acid sequence of the growth hormone (somatotropic hormone) is also known; it seems to stimulate the synthesis of RNA and in this way to accelerate growth. The luteinizing hormone, a mucoprotein containing about 12 percent carbohydrate, consists of two subunits, each with a molecular weight of approximately 15,000; when separated, the subunits recombine spontaneously. The urine of pregnant women contains chorionic gonadotropin, the presence of which makes possible early diagnosis of pregnancy. The amino acid sequence is known. The sequence of 160 of its 190 amino acids is identical with those of the growth hormone; 100 of these also occur in the same sequence as in lactogenic hormone. The different pituitary hormones and the chorionic gonadotropin thus may have been derived from a common substance that, during evolution, underwent differentiation.

Peptides with Hormonelike Activity

Small peptides have been discovered that, like hormones, act on certain target organs. One peptide, angiotensin (angiotonin or hypertensin), is formed in the blood from angiotensinogen by the action of renin, an enzyme of the kidney. It is an octapeptide and increases blood pressure. Similar peptides include bradykinin, which stimulates smooth muscles; gastrin, which stimulates secretion of hydrochloric acid and pepsin in the stomach; secretin, which stimulates the flow of pancreatic juice; and kallikrein, the activity of which is similar to bradykinin.

Protein Metabolism

Protein metabolism is the chemical cycle of breaking down protein (catabolism) and using the components to synthesizing (anabolism) new molecules to be used in the body. The process is also known as proteometabolism.

Proteins, fats, and carbohydrates (called macronutrients) are part of a complex metabolic cycle that is essential to life. During digestion food containing these nutrients is chemically broken down into its basic components and absorbed for use in the body. Protein molecules are split into their basic building blocks, called amino acids, which are then chemically re-arranged to synthesize new proteins that the body needs. Fats are broken down into fatty acids and cholesterol, and carbohydrates are split into simple sugars such as glucose and fructose, which provide most of the energy to drive chemical reactions in the body. These smaller, simpler molecules are absorbed in the small intestine and enter the circulatory system. They then pass through the liver where some of these "building block molecules" are synthesized into more complex compounds needed by the body.

In order for cellular metabolism to occur, two reactions happen continuously. Small molecules are build up into larger molecules, a process called anabolism or constructive metabolism, while large molecules are broken down into their component parts during a process called catabolism or destructive metabolism. This building up and breaking down is regulated by a complex set of hormones and enzymes, themselves proteins. For an individual to remain healthy, the processes of anabolism and catabolism must remain in balance.

Catabolism, or the breakdown of nutrients obtained from food, releases energy that drives all metabolic activities in the body. For example, glucose is broken down to provide energy for cellular respiration that allows functions such as muscle movement. Proteins are broken into amino acids then re-synthesized into hormones and enzymes to regulate chemical reactions in the cell, and molecules used for tissue growth and repair. Carbohydrates and fats are the preferred sources of energy for cellular metabolism. When the supply of fats and carbohydrates is insufficient to meet the body' needs, proteins can be broken down to supply energy. This accounts for the loss of muscle seen in prolonged cases of starvation.

Protein is a nitrogen-containing compound found in all plants and animals. There is a continuous need for protein to make hormones, enzymes, antibodies, and to produce new tissue (growth) and repair damaged tissues (maintenance). About 75% of human body tissue is made of protein. From 1-2% of the body's total protein is broken down each day into amino acid and recycled into new proteins. About60-70% of the amino acids the body needs come from this recycling process.

Digestion of Protein

The digestion of proteins begins in the stomach where the hormone pepsin is secreted by the stomach. Pepsin breaks the long polypeptide molecules into smaller peptides. The mechanical churning of the stomach assists digestion by mixing food with gastric (stomach) secretions. When the contents of the stomach reach a certain degree of acidity, the pyloric sphincter, a muscle that separates the stomach from the small intestine, opens. The stomach contents, called chyme, flow into the duodenum, or upper part of the small intestine. As chyme moves through the small intestine, enzymes break the chemical bonds in the peptides, reducing the proteins to their component amino acids. (Breakdown of fats and carbohydrates is occurring simultaneously under the direction of different hormones and enzymes). In the small intestine, intestinal cells and blood vessels are in close proximity, separated only by cell membranes. Amino acids molecules are small enough that they can move though the intestinal wall and into blood where they (along with glucose, fatty acids, and the other products of digestion) are carried by a large blood vessel to the liver.

The liver is at the heart of protein metabolism. It has both anabolic and catabolic functions. In the liver, amino acids are synthesized into larger proteins that circulate through the body performing a huge variety of tasks including stimulating production of other proteins. The liver also breaks down proteins, for example, hemoglobin found in dead red blood cells. The liver cleanses the blood by removing cellular debris and processing excess nitrogen that is produced by chemical reactions within cells. This excess nitrogen is initially in the form of ammonia (NH3. If allowed to remain in the blood, it would rapidly become toxic to the body. The liver converts ammonia to non-toxic urea that is removed from the body in urine. This lost nitrogen must then be replaced through diet.

Function

Protein metabolism consists of a cycle of breaking down proteins, synthesizing new ones and removing nitrogenous waste products that result from these reactions. The amount of protein needed to balance this cycle changes throughout an individual' life. Growing children who are creating new muscle and bone, for example, have higher protein needs than adults.

Protein Synthesis

The overall process of protein synthesis extends from gene transcription in the nucleus to polypeptide synthesis on ribosomes in the cytoplasm and is summarized in figure.

Genes and the Genetic Code

The human genome contains approximately 20,000 protein-coding genes that provide the instructions for the 250,000 to 1,000,000 proteins that operate in the human body over its lifetime. The number of proteins exceeds the number of genes because one gene can code for more than one protein and many proteins exist in multiple forms due to post-synthetic chemical modifications. In addition there are thousands of non-coding RNA genes that help regulate the protein coding genes.

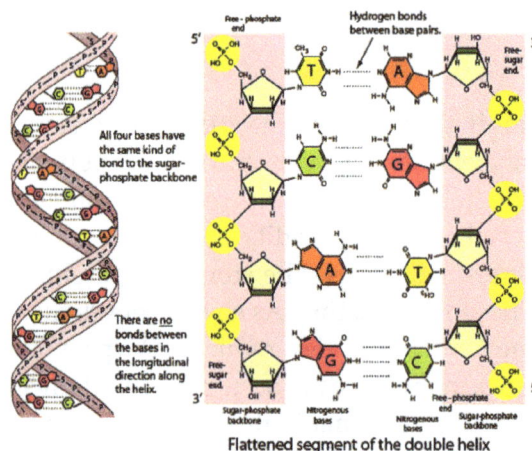

Figure: DNA structure showing the double helix and the base pairing (hydrogen bonds)
where A pairs with T and G pairs with C.

Genetic information is stored in the nucleus of cells in the form of deoxyribose nucleic acid (DNA). DNA is composed of just four building blocks (bases) adenine (A), guanine (G), cytosine (C) and thymine (T) linked by a deoxyribose -phosphate backbone to form a double helix.

Second Letter

		U		C		A		G		
1st letter	**U**	UUU UUC	Phe	UCU UCC	Ser	UAU UAC	Tyr	UGU UGC	Cys	U C
		UUA UUG	Leu	UCA UCG		UAA UAG	Stop Stop	UGA UGG	Stop Trp	A G
	C	CUU CUC CUA CUG	Leu	CCU CCC CCA CCG	Pro	CAU CAC	His	CGU CGC CGA CGG	Arg	U C A G
						CAA CAG	Gln			
	A	AUU AUC AUA	Ile	ACU ACC ACA ACG	Thr	AAU AAC	Asn	AGU AGC	Ser	U C A G
		AUG	Met			AAA AAG	Lys	AGA AGG	Arg	
	G	GUU GUC GUA GUG	Val	GCU GCC GCA GCG	Ala	GAU GAC	Asp	GGU GGC GGA GGG	Gly	U C A G
						GAA GAG	Glu			**3rd letter**

Figure: The genetic code.

The genetic sequence of DNA bases A, G, C, and T is copied and processed into the corresponding sequence of messenger RNA bases A, G, C and U for translation into polypeptide by the ribosome. The sequence of bases is read in triplets with sixty four possible combinations. Methionine (Met) and tryptophan (Trp) are each coded for by a single triplet (codon), asparagine (Asn), aspartic acid (Asp), glutamine (Gln), glutamic acid (Glu), cysteine (Cys), phenylalanine (Phe), tyrosine (Tyr) and lysine (Lys) each have two codons; isoleucine (Ile) has three; glycine (Gly), alanine (Ala), valine (Val), threonine (Thr) and proline (Pro) have four codons while serine (Ser) and arginine (Arg) have six. There are three stop codons that denote the C-terminus of the translated protein.

As there are 20 amino acids, the code is not read as a single letter (4 possibilities only) or double letter (4x4 – 16 combinations) format, rather it is read in triplets called codons, with 4x4x4 (64) total combinations coding for all 20 amino acids as well as some punctuation (Stop/Start) instructions.

Transcription

Proteins are not synthesized directly from DNA, but from an RNA (ribose nucleic acid) copy derived from one strand of DNA by a process called transcription. Transcription occurs in the nucleus. The sections of the RNA gene copy that correspond to regions in the DNA that do not code for residues in the final protein (introns), are removed and the processed RNA copy - called messenger RNA - is transported to the cytoplasm where protein synthesis takes place. RNA contains three of the same bases as DNA, (adenine, guanine, cytosine) but employs uracil as the fourth base rather than thymine. Gene transcription is controlled by special DNA-binding proteins called transcription factors that are synthesized and activated/inactivated by protein hormones such as insulin and glucagon and non-protein hormones such as corticosteroids.

Role of Transfer RNA

A transfer RNA (tRNA) is an adaptor molecule composed of RNA, typically 76 to 90 nucleotides in length. It serves as the physical link between the nucleotide sequence of the mRNA being translated and the resulting amino acid sequence of the synthesized protein. Thus tRNA is the means by which the correct amino acid - required to match each codon (triplet of bases) in the transcribed mRNA- is positioned in the peptidyl-transferase centre of the ribosome. One end of each tRNA contains a three-nucleotide sequence called the anticodon that can form three base pairs with a complementary three-nucleotide codon in mRNA during protein biosynthesis. Covalently attached to the other end of each tRNA is the amino acid that corresponds to the mRNA codon sequence. Each type of tRNA molecule can have only one type of amino acid attached to it and this is synthesized by enzymes called *aminoacyl tRNA synthetases*. One molecule of ATP is consumed in this process. Given the genetic code contains multiple codons that specify the same amino acid there will be several tRNA molecules bearing different anticodons which also carry the same amino acid.

Translation on Ribosomes

Protein synthesis is carried out in the cytoplasm by ribosomes - massive protein and RNA complexes that translate the nucleotide code on messenger RNA (mRNA) into functional protein. Eukaryotic organisms, which include humans, have two ribosomal subunits, the large 60S and small 40S, which combine to form the functional 80S complex. In contrast, prokaryotes such as bacteria have similar, but smaller subunits — a large 50S and small 30S, which combine to form a 70S complex.

Ribosomes have been the focus of structural and biochemical studies for more than 50 years and in 2000, Tom Steitz's laboratory at Yale University in Hartford Connecticut, published a high-resolution (2.4 Å) structure, of the large 50S subunit in the journal Science. At this resolution, the researchers were able to definitively place nearly all of the 50S subunit's 3,045 nucleotides and 31 proteins. The structure revealed that the ribosome is a ribozyme because the catalytic *peptidyl transferase* activity that catalyses peptide bond formation, linking the amino acids together in the growing peptide chain, is performed by the ribosomal RNA. Numerous initiation, elongation and release factors ensure that protein synthesis occurs progressively and with high specificity. In the past few years, high-resolution structures have provided molecular snapshots of different intermediates in ribosome-mediated translation in atomic detail. Together, these studies have revolutionized our understanding of the mechanism of protein synthesis.

The process of protein synthesis on ribosomes involves binding the mRNA in a tunnel formed between the two ribosomal subunits and initiating protein synthesis at the first codon. The two ribosomal subunits perform different roles in protein synthesis. The small ribosomal subunit mediates the correct inter¬actions between the anticodons of the tRNAs and the codons in the mRNA that they are translating in order to determine the order of the amino acids in the protein being synthesized. The large subunit contains the *peptidyl-transferase* centre (PTC), which catalyses the formation of peptide bonds in the growing polypeptide.

Both subunits contain three binding sites A, P and E, for tRNA molecules that are in three different functional states. The A site binds the aminoacyl-tRNA that is about to be incorporated into the growing polypeptide chain, the P site positions the peptidyl-tRNA (ie the tRNA with the growing peptide chain attached) and the E site is occupied by the deacylated tRNA before it dissociates from the ribosome (ie the tRNA after its attached peptide chain has been transferred (covalently linked) to the incoming amino acid on the aminoacyl tRNA).

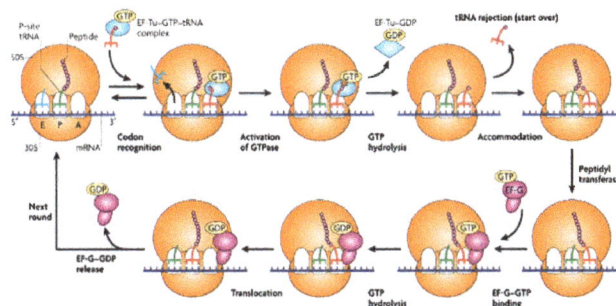

Figure: An overview of steps in protein synthesis.

mRNA translation is initiated with the binding of tRNAfmet to the P site (not shown). An incoming tRNA is delivered to the A site in complex with elongation factor (EF)-Tu–GTP. Correct codon–anticodon pairing activates the GTPase centre of the ribosome, which causes hydrolysis of GTP and release of the aminoacyl end of the tRNA from EF Tu. Binding of tRNA also induces conformational changes in ribosomal (r)RNA that optimally orientates the peptidyl-tRNA and aminoacyl-tRNA for the peptidyl-transferase reaction to occur, which involves the transfer of the peptide chain onto the A site tRNA. The ribosome must then shift in the 3′ mRNA direction so that it can decode the next mRNA codon. Translocation of the tRNAs and mRNA is facilitated by binding of the GTPase EF G, which causes the deacylated tRNA at the P site to move to the E site and the peptidyl-tRNA at the A site to move to the P site upon GTP hydrolysis. The ribosome is then ready for the next round of elongation. The deacylated tRNA in the E site is released on binding of the next aminoacyl-tRNA to the A site. Elongation ends when a stop codon is reached, which initiates the termination reaction that releases the polypeptide. Reprinted by permission from Macmillan Publishers Ltd:

Of central interest are the mechanisms of peptide bond formation and mRNA decoding, which are crucial processes in the elongation phase of protein synthesis by the ribosome. During this phase of protein synthesis, nascent polypeptides are elongated from the N- to the C-terminus by the addition of one amino acid at a time. This process is facilitated by two protein factors: elongation factor Tu (EF-Tu), which facilitates the delivery of aminoacyl-tRNA to the A site of the ribosome, and elongation factor G (EF G), which promotes the translocation of the tRNAs and associated mRNA from their positions in the A and P sites to the P and E sites, respectively, and dissociates the previously bound E-site tRNA.

The accurate delivery of the correct aminoacyl-tRNA to the A site involves at least two distinct steps: (i) an interaction between the anticodon base triplet in the tRNA and the corresponding codon of the mRNA that resides in the A site of the ribosome and (ii) the communication of this correct formation of anticodon/codon Watson–Crick base pairing to the GTPase centre located in the large ribosomal subunit ~70 Å away, which results in the hydrolysis of the GTP bound to EF Tu. This GTP hydrolysis changes the conformation of EF-Tu resulting in its release from the tRNA and ribosome and the subsequent accommodation of the aminoacyl end of the tRNA into the peptidyl-transferase centre (PTC), which is followed rapidly by peptide bond formation.

At the end of the elongation cycle when the stop codon has been positioned in the A site, one of two protein release factors (RFs), RFI or RFII, binds to the A site and promotes the deacylation of the peptidyl-tRNA. A recycling factor, with the help of EF G, then leads to the dissociation of the release factor and the two ribosomal subunits.

Protein Targeting

A typical mammalian cell may contain numerous kinds of proteins and numerous individual protein molecules. The eukaryotic cell is a multi-compartmental structure. Its many organelles each requires different proteins. Except a few of them which are synthesized in mitochondria and chloroplasts all other proteins necessary for the cell and the ones to be secreted by the cell are synthesized in the cytosol on free ribosomes and on ribosomes bound to the endoplasmic reticulum.

Most proteins are coded by the nuclear genome and synthesized in the cytoplasm. The proteins are present in the ER, mitochondria, chloroplasts, Golgi, peroxisomes, nucleus, in the cytosol and in the membranes of all these organelles. They are selectively transported into their appropriate organelles inside the cell and across the plasma membrane to be secreted outside the cell.

Some of them are carried into membrane bound vesicles which bud off from one organelle and transported in definite pathways. Different destinations of different proteins require sophisticated system for labelling and sorting newly synthesized proteins and ensuring that they reach their proper places. This transportation of proteins to their final destinations is called protein targeting.

Proteins destined for cytoplasm and those to be incorporated into mitochondria, chloroplasts and nuclei are synthesized on free ribosomes in the cytoplasm. Proteins destined for cellular membranes, lysosomes and extracellular transport, use a special distribution system. The main structures in this system are the rough endoplasmic reticulum (RER) and Golgi complex.

The RER is a network of interconnected membrane enclosed vesicles or vacuoles. The endoplasmic reticulum is coated with polyribosomes to give it a rough appearance. The golgi complex is also a stack of membrane bound sacs but they are not interconnected. The golgi complex acts as a switching center for proteins to various destinations.

Proteins to be directed to their destinations via Golgi complex are synthesized by ribosomes associated with endoplasmic reticulum.

Protein sorting requires proper address labels which are in the form of peptide signal sequences. A signal sequence that directs the protein to its target is present in the form of 13-35 amino acids in the newly synthesized protein itself. It is the first to be synthesized and is mostly present at the amino N-terminal, sometimes at the carboxyl C- terminal.

It is known as signal sequence or leader sequence. Some proteins are further sorted to a sub-compartment within the target organelle. For this purpose, a second signal sequence is present behind the first signal sequence which is cleaved.

Proteins carried inside the membrane bound vesicles are called cargo proteins. An embedded or integrated protein is carried in the membrane of the vesicle, while secretory protein is carried within the lumen of the vesicle. The vesicle buds off from the donor surface and fuses with the target surface releasing its contents into the target organelle and the membrane protein is incorporated into the membrane of the target organelle. The process is repeated during the passage of protein from ER to Golgi to lysosomes and from Golgi to plasma membrane.

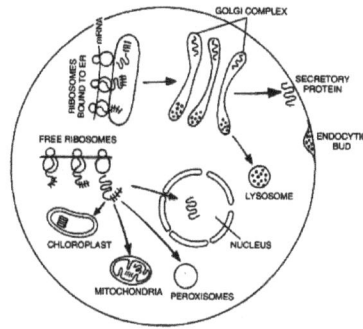

Figure: Nascent proteins targeting to different organelles of the cell and cell secretion.

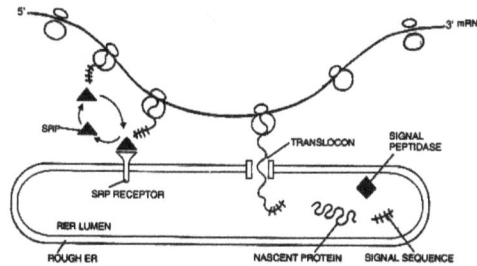

Figure: Transport of proteins into ER.

Transport of Proteins into ER

A short N-terminus signal sequence at the beginning of the growing nascent protein chain' determines whether a ribosome synthesizing the proteins binds to ER or not. The protein synthesis always begins on free ribosomes. As the signal sequence emerges out of the ribosome, the large ribosomal sub-unit binds to ER membrane.

This is decided by the type of signal sequence. This is the first sorting as the ribosome binds to ER, forming rough ER. Translocation takes place into the ER while growing chain is still bound to the ribosome. This is called co-translational translocation. The process is facilitated by the signal sequence recognition mechanism.

Signal Sequence Recognition Mechanism

It consists of a signal recognition particle (SRP) present in the cytosol. SRP binds to the signal sequence of the nascent protein as soon as it emerges out of ribosome and directs it towards the ER membrane. The binding of SRP stops further synthesis of protein chain when it is about 70 amino acids long.

This prevents it from folding. The SRP-ribosome complex binds to the SAP receptor, which is an integral membrane protein in the wall of ER and is a docking protein of the ER. At this point GTP hydrolysis hydrolyses frees SRP which is ready for the next round of directing next nascent protein of ER.

Now lengthening of nascent polypeptide restarts which enters ER lumen. Ribosome is aligned to a channel in the wall of ER. This channel is called translocon. It allows the elongating chain to enter the translocon into the ER lumen.

As the growing polypeptide chain emerges into the ER lumen, the signal sequence is cleaved by a peptide called signal peptidase. Inside the lumen, the protein may become folded into its final active form or may be carried into its secretary pathway or may be embedded in the ER membrane.

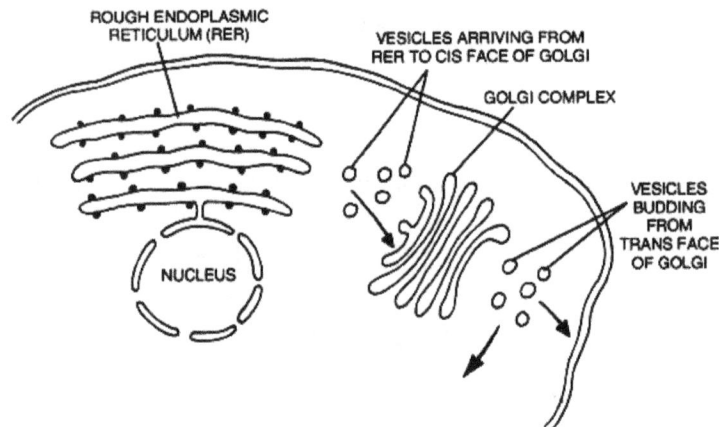

Figure: Transfer of proteins from RER to Golgi Complex

Once inside the lumen of ER, the protein undergoes folding and several modifications for which the ER lumen contains a number of enzymes and chaprone proteins. The most common processing is glycosylation which involves addition of carbohydrates to the protein chain. Glycosylation generally occurs in the ER lumen but sometimes in Golgi also.

Most oligosachharides or glycons are attached to the amino group NH3 and the proteins are called N-linked glycoprotiens e.g. oligosachharide attached to aspargine. A preformed oligosachharide is added to the proteins. This structure is Man 9 (Glc NAC)$_2$ called high mannose structure.

This contains mannose, glucose and N-acetylglucosamine). All nascent proteins start the sorting pathway by addition of the same pre-formed oligosachharide in plants and animals. Almost all proteins that enter the secretary pathway are glycosylated.

In ER lumen, after glycosylation, many protiens are folded and stabilized by disulphide proteins bonds (-S-S-). This reaction is catalyzed by an enzyme, protein disulphide isomerase (PDI). Most of human proteins are stabilized by disulphide bonds.

Role of Golgi Complex in Protein Transportation

The role of Golgi complex is to act as a switching center for proteins to various destinations. Both ER and Golgi apparatus are flattened cisternae. Transport of proteins from one compartment (donor) to the next one (target) is carried out in transport vesicles. The vesicles contain cargo proteins in their lumen and integral membrane proteins in their membranes.

The vesicles bud off from ER and fuse with the cis-compartment or receiving compartment of Golgi. In this process cargo proteins are delivered into the lumen of Golgi and membrane proteins become part of the membrane of the target vesicles. The proteins are glycosylated, folded, modified and sorted in ER. This process of glycosylation, modification and sorting of proteins continues in successive Golgi cisternae.

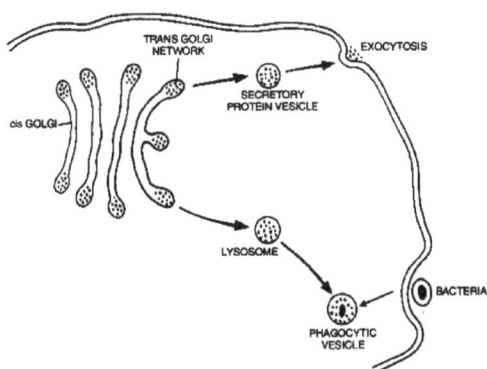

Figure: Transport of proteins from Golgi to Lysosomes.

Starting from the cis-compartment to medial compartment and lastly to trans-Golgi network proteins are exported to the end target. In trans-golgi network (TGN) proteins are further sorted to be delivered to lysosomes, for secretion outside the cell and to plasma membrane according to signals present in the nascent proteins.

Transport of Proteins from Golgi to Lysosomes

The lysosomal enzymes and lysosomal membrane proteins are synthesized in rough ER and transported to Golgi cisternae and ultimately to lysosomes. The sorting signal that directs the lysosomal enzymes from the trans- Golgi network (TGN) to lysosomes is mannose 6-phosphate (M6P). The attachment of M6P to lysosomal enzymes prevents their further modification.

Separation of M6P bearing lysosomal enzymes from other proteins takes place in TGN. The wall of TGN contains M6P receptors. These M6P receptors bind to lysosomal proteins. The vesicles containing these receptor bearing proteins bud off from TGN. These vesicles are called lysosomes. Later these vesicles fuse with vesicles which have arisen by pinacocytosis and phagocysis to form secondary lysosomes. Low pH of Lysosomes triggers the dissociation of enzymes from the receptors.

The M6P receptors are recycled back to trans-golgi network in vesicles. Lysosomes contain hydrolyzing proteolytic enzyme, which digests proteins meant for degradation. A protein named ubiquitin marks the proteins meant for destruction. Ubiquitin is present in all eukaryotic cells. This mechanism degrades only those proteins which are meant for destruction and not the proteins which are to be left alone.

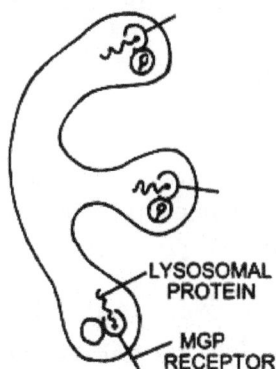

Figure: Trans-Gologi Network.

The proteins meant for secretion travel to plasma membrane from trans-golgi network.

All this transportation of vesicles from the RER to the cis face of golgi to successive levels of golgi and on to their final destinations requires the high levels of specificity in targeting. Transport of vesicles to wrong destinations would lead to cellular chaos.

Targeting of Proteins to Mitochondria and Chloroplasts

Mitochondria and chloroplasts possess their own DNA, ribosomes, mRNA and synthesize a few proteins. But most of the proteins required for mitochondria and chloroplasts are synthesized in cytosol by nuclear DNA and then imported into these organelles. Both these organelles are covered by double membranes. The proteins are translocated into these organelles after they are fully synthesized. This is known as post-translational translocation.

There are four mitochondrial locations where the proteins are targeted. These are outer membrane, inner membrane, intermembranal space and mitochondrial matrix. The proteins are released in unfolded state and they bind to a family of chaprones. These chaprones are cytosolic hsp 70 proteins (heat shock proteins) that deliver the proteins to an import receptor on the outer mitochondrial membrane.

The import receptor then slides to a site where inner and otuer mitochondrial membrane form a channel through which the unfolded protein enters into mitochondria leaving out cytosolic hsp 70 protiens. As the protein reaches the matrix, mitochondrial heat shock protein, mitochondrial hsp 70 binds to it. A protease cleaves the signal sequence.

These proteins have more than one successive N-terminal targeting signal sequence. The first signal sequence imports the protein into matrix and the second signal re-directs the protein into membranes or inter membranal space.

Mitochondria processes machinery for cellular respiration. Each membrane and each compartment of mitochondria has its unique proteins. Enzymes of electron transport chain lie in the inner membrane while most enzymes of citric acid cycle are found in the matrix.

Protein Targeting to Chloroplasts

The newly synthesized proteins by free ribosomes are impored into chloroplasts as in mitochondria. Calvin cycle enzymes fix atmospheric CO_2 into carbohydrates during photosynthesis.

Protein Targeting into Nucleus

The nuclear envelope consists of outer and inner membranes and has inter membranous space between them. The outer membrane is continuous with ER and has ribosomes on it. Proteins for the nucleus are synthesized on free ribosomes in the cytosol and imported into nucleus through 3000-4000 nuclear pores known as nuclear pore complexes which are special gates.

The proteins that are imported into nucleus are in fully folded state and do not require any chaprones. Protiens imported into nucleus have targeting signal sequences on them which are called nuclear localization signals (NLS). Each one has 4-8 amino acids and they are internal sequences and not terminal. NLS is not cleaved from the protein. Due to this feature proteins can re-enter the nucleus whenever the nuclear envelope is lost during cell division.

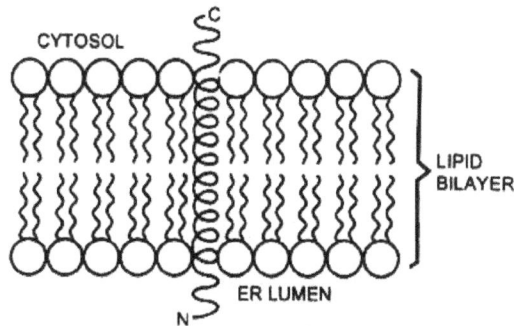

Figure: single pass topogenic sequence.

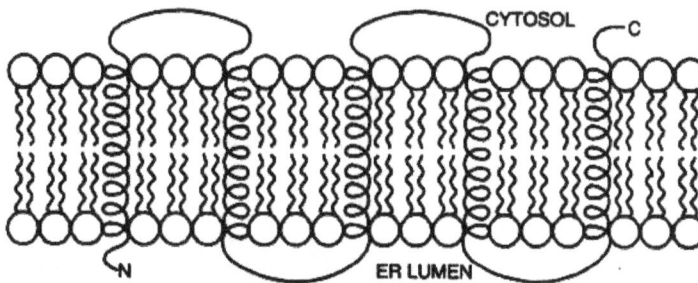

Figure: Multi pass protein domains across ER membrane.

Membrane Proteins

The proteins embedded in different membranes may have single trans-membrane domain which is a segment of 20-25 amino acids. Other proteins may have many trans-membrane domains connected by loops on both sides of the membrane. These proteins are called multi-pass orientation proteins. In photosynthetic bacteria a protein called bacterio-rodospin spans 12-14 times across the lipid bilayer membrane of bacteria. It traps energy from sunlight and uses it to pump protons across the bacterial membrane.

Protein Glycosylation

The addition of a carbohydrate moiety to a protein molecule is referred to as protein glycosylation. It is a common post translational modification for protein molecules involved in cell membrane formation. During this process, the linking of monosaccharide units to the amino acid chains sets up the stage for a series of enzymatic reactions that lead to the formation of glycoproteins (n and o linked oligosaccharides that are found to a protein entity). In all 16 known enzymes are supposed to mediate this reaction. A typical glycoprotein has at least 41 bonds which involve 8 amino acids and 13 different monosaccharide units and includes the glycophosphatidylinositol (GPI) and phosphoglycosyl linkages. Protein glycosylation helps in proper folding of proteins, stability and in cell to cell adhesion commonly needed by cells of the immune system. The major sites of protein glycosylation in the body are ER, Golgi body, nucleus and the cell fluid.

(N-linkage)

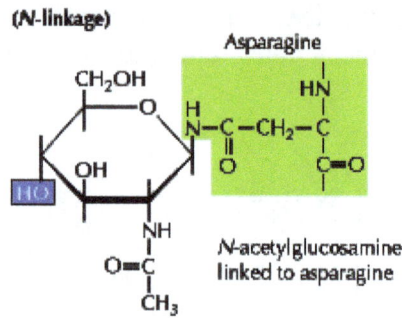

N-acetylglucosamine linked to asparagine

Protein glycosylation can be categorized in two main types:

a) N-Linked glycosylation in proteinsa) N-linked glycosylation: It begins with the addition of a 14-sugar precursor to an asparagine amino acid. It contains glucose, mannose and n-acetylglucos-amine molecules. This entity is then transferred to the ER lumen. The oligosaccharyl transferase enzyme attaches the oligosaccharide chain to asparagine that occurs in the tripeptide sequence, Asn-X-Ser or Asn-X-Thr. X can be any amino acid other than Proline. The oligosaccharide attached protein sequence now folds correctly and is now translocated to the Golgi body where the mannose residue is removed.

(O-linkage)

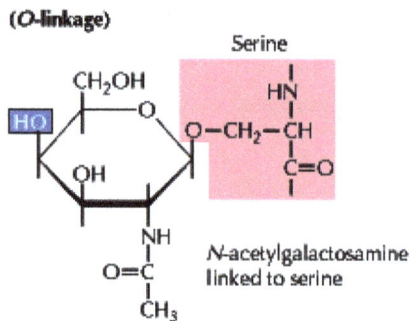

N-acetylgalactosamine linked to serine

b) O-linked glycosylation: Glycosylation begins with an enzyme mediated addition of N-acetyl-galactosamine followed by other carbohydrates to serine or threonine residues.

References

- Plant-proteins, protein, science: britannica.com , Retrieved August 9, 2019

- Protein-metabolism, encyclopedias-almanacs-transcripts-and-maps, medicine: encyclopedia.com, Retrieved June 7, 2019.

- Protein-metabolism, metabolism-and-hormones: diapedia.org , Retrieved February 15, 2019

- Protein-targeting-with-diagram-molecular-biology, protein-targeting, proteins: biologydiscussion.com , Retrieved May 21, 2019

- Glycosylation, glossary, glycan: premierbiosoft.com , Retrieved June 19, 2019

<h1>Chapter 5</h1>

<h1>Nucleic Acid</h1>

The overall name for DNA and RNA is known as nucleic acids. They are the small biomolecules that are composed of nucleotides which are made up of three components: 5-a carbon sugar, nitrogenous base and a phosphate group. The topics elaborated in this chapter will help in gaining a better perspective about the nucleic acids.

Nucleic Acid is a naturally occurring chemical compound that is capable of being broken down to yield phosphoric acid, sugars, and a mixture of organic bases (purines and pyrimidines). Nucleic acids are the main information-carrying molecules of the cell, and, by directing the process of protein synthesis, they determine the inherited characteristics of every living thing. The two main classes of nucleic acids are deoxyribonucleic acid (DNA) and ribonucleic acid (RNA). DNA is the master blueprint for life and constitutes the genetic material in all free-living organisms and most viruses. RNA is the genetic material of certain viruses, but it is also found in all living cells, where it plays an important role in certain processes such as the making of proteins.

Nucleotides: Building Blocks of Nucleic Acids

Basic Structure

Nucleic acids are polynucleotides—that is, long chainlike molecules composed of a series of nearly identical building blocks called nucleotides. Each nucleotide consists of a nitrogen-containing aromatic base attached to a pentose (five-carbon) sugar, which is in turn attached to a phosphate group. Each nucleic acid contains four of five possible nitrogen-containing bases: adenine (A), guanine (G), cytosine (C), thymine (T), and uracil (U). A and G are categorized as purines, and C, T, and U are collectively called pyrimidines. All nucleic acids contain the bases A, C, and G; T, however, is found only in DNA, while U is found in RNA. The pentose sugar in DNA (2′-deoxyribose) differs from the sugar in RNA (ribose) by the absence of a hydroxyl group (−OH) on the 2′ carbon of the sugar ring. Without an attached phosphate group, the sugar attached to one of the bases is known as a nucleoside. The phosphate group connects successive sugar residues by bridging the 5′-hydroxyl group on one sugar to the 3′-hydroxyl group of the next sugar in the chain. These nucleoside linkages are called phosphodiester bonds and are the same in RNA and DNA.

Biosynthesis and Degradation

Nucleotides are synthesized from readily available precursors in the cell. The ribose phosphate portion of both purine and pyrimidine nucleotides is synthesized from glucose via the pentose phosphate pathway. The six-atom pyrimidine ring is synthesized first and subsequently attached to the ribose phosphate. The two rings in purines are synthesized while attached to the ribose phosphate during the assembly of adenine or guanine nucleosides. In both cases the end product is a nucleotide carrying a phosphate attached to the 5′ carbon on the sugar. Finally, a specialized enzyme called a kinase adds two phosphate groups using adenosine triphosphate (ATP) as the phosphate donor to form ribonucleoside triphosphate, the immediate precursor of RNA. For DNA, the 2'-hydroxyl group is removed from the ribonucleoside diphosphate to give deoxyribonucleoside diphosphate. An additional phosphate group from ATP is then added by another kinase to form a deoxyribonucleoside triphosphate, the immediate precursor of DNA.

During normal cell metabolism, RNA is constantly being made and broken down. The purine and pyrimidine residues are reused by several salvage pathways to make more genetic material. Purine is salvaged in the form of the corresponding nucleotide, whereas pyrimidine is salvaged as the nucleoside.

DNA

Deoxyribonucleic acid, or DNA, is a molecule that contains the instructions an organism needs to develop, live and reproduce. These instructions are found inside every cell, and are passed down from parents to their children.

DNA Structure

DNA is made up of molecules called nucleotides. Each nucleotide contains a phosphate group, a sugar group and a nitrogen base. The four types of nitrogen bases are adenine (A), thymine (T), guanine (G) and cytosine (C). The order of these bases is what determines DNA's instructions, or genetic code.

Similar to the way the order of letters in the alphabet can be used to form a word, the order of nitrogen bases in a DNA sequence forms genes, which in the language of the cell, tells cells how to make proteins. Another type of nucleic acid, ribonucleic acid, or RNA, translates genetic information from DNA into proteins.

Nucleotides are attached together to form two long strands that spiral to create a structure called a double helix. The bases on one strand pair with the bases on another strand: adenine pairs with thymine, and guanine pairs with cytosine.

DNA molecules are long — so long, in fact, that they can't fit into cells without the right packaging. To fit inside cells, DNA is coiled tightly to form structures called chromosomes. Each chromosome contains a single DNA molecule. Humans have 23 pairs of chromosomes, which are found inside the cell's nucleus.

DNA Replication in Bacteria

DNA replication has been well studied in bacteria primarily because of the small size of the genome and the mutants that are available. E. coli has 4.6 million base pairs (Mbp) in a single circular chromosome and all of it is replicated in approximately 42 minutes, starting from a single origin of replication and proceeding around the circle bidirectionally (i.e., in both directions). This means that approximately 1000 nucleotides are added per second. The process is quite rapid and occurs with few errors.

DNA replication uses a large number of proteins and enzymes (Table).

Table: The Molecular Machinery Involved in Bacterial DNA Replication	
Enzyme or Factor	Function
DNA pol I	Exonuclease activity removes RNA primer and replaces it with newly synthesized DNA
DNA pol III	Main enzyme that adds nucleotides in the 5' to 3' direction
Helicase	Opens the DNA helix by breaking hydrogen bonds between the nitrogenous bases
Ligase	Seals the gaps between the Okazaki fragments on the lagging strand to create one continuous DNA strand
Primase	Synthesizes RNA primers needed to start replication
Single-stranded binding proteins	Bind to single-stranded DNA to prevent hydrogen bonding between DNA strands, reforming double-stranded DNA
Sliding clamp	Helps hold DNA pol III in place when nucleotides are being added
Topoisomerase II (DNA gyrase)	Relaxes supercoiled chromosome to make DNA more accessible for the initiation of replication; helps relieve the stress on DNA when unwinding, by causing breaks and then resealing the DNA
Topoisomerase IV	Introduces single-stranded break into concatenated chromosomes to release them from each other, and then reseals the DNA

One of the key players is the enzyme DNA polymerase, also known as DNA pol. In bacteria, three main types of DNA polymerases are known: DNA pol I, DNA pol II, and DNA pol III. It is now known that DNA pol III is the enzyme required for DNA synthesis; DNA pol I and DNA pol II are primarily required for repair. DNA pol III adds deoxyribonucleotides each complementary to a nucleotide on the template strand, one by one to the 3'-OH group of the growing DNA chain. The addition of these nucleotides requires energy. This energy is present in the bonds of three phosphate groups attached to each nucleotide (a triphosphate nucleotide), similar to how energy is stored in the phosphate bonds of adenosine triphosphate (ATP). When the bond between the phosphates is broken and diphosphate is released, the energy released allows for the formation of a covalent phosphodiester bond by dehydration synthesis between the incoming nucleotide and the free 3'-OH group on the growing DNA strand.

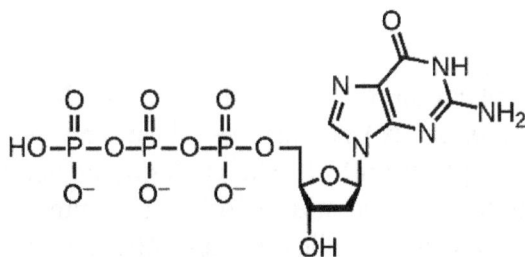

Figure: This structure shows the guanosine triphosphate deoxyribonucleotide that is incorporated into a growing DNA strand by cleaving the two end phosphate groups from the molecule and transferring the energy to the sugar phosphate bond. The other three nucleotides form analogous structures.

Initiation

The initiation of replication occurs at specific nucleotide sequence called the origin of replication, where various proteins bind to begin the replication process. E. coli has a single origin of replication (as do most prokaryotes), called oriC, on its one chromosome. The origin of replication is approximately 245 base pairs long and is rich in adenine-thymine (AT) sequences.

Some of the proteins that bind to the origin of replication are important in making single-stranded regions of DNA accessible for replication. Chromosomal DNA is typically wrapped around histones (in eukaryotes and archaea) or histone-like proteins (in bacteria), and is supercoiled, or extensively wrapped and twisted on itself. This packaging makes the information in the DNA molecule inaccessible. However, enzymes called topoisomerases change the shape and supercoiling of the chromosome. For bacterial DNA replication to begin, the supercoiled chromosome is relaxed by topoisomerase II, also called DNA gyrase. An enzyme called helicase then separates the DNA strands by breaking the hydrogen bonds between the nitrogenous base pairs. Recall that AT sequences have fewer hydrogen bonds and, hence, have weaker interactions than guanine-cytosine (GC) sequences. These enzymes require ATP hydrolysis. As the DNA opens up, Y-shaped structures called replication forks are formed. Two replication forks are formed at the origin of replication, allowing for bidirectional replication and formation of a structure that looks like a bubble when viewed with a transmission electron microscope; as a result, this structure is called a replication bubble. The DNA near each replication fork is coated with single-stranded binding proteins to prevent the single-stranded DNA from rewinding into a double helix.

Once single-stranded DNA is accessible at the origin of replication, DNA replication can begin. However, DNA pol III is able to add nucleotides only in the 5′ to 3′ direction (a new DNA strand can be only extended in this direction). This is because DNA polymerase requires a free 3′-OH group to which it can add nucleotides by forming a covalent phosphodiester bond between the 3′-OH end and the 5′ phosphate of the next nucleotide. This also means that it cannot add nucleotides if a free 3′-OH group is not available, which is the case for a single strand of DNA. The problem is solved with the help of an RNA sequence that provides the free 3′-OH end. Because this sequence allows the start of DNA synthesis, it is appropriately called the primer. The primer is five to 10 nucleotides long and complementary to the parental or template DNA. It is synthesized by RNA primase, which is an RNA polymerase. Unlike DNA polymerases, RNA polymerases do not need a free 3′-OH group to synthesize an RNA molecule. Now that the primer provides the free 3′-OH group, DNA polymerase III can now extend this RNA primer, adding DNA nucleotides one by one that are complementary to the template strand.

Elongation

During elongation in DNA replication, the addition of nucleotides occurs at its maximal rate of about 1000 nucleotides per second. DNA polymerase III can only extend in the 5′ to 3′ direction, which poses a problem at the replication fork. The DNA double helix is antiparallel; that is, one strand is oriented in the 5′ to 3′ direction and the other is oriented in the 3′ to 5′ direction (see Structure and Function of DNA). During replication, one strand, which is complementary to the 3′ to 5′ parental DNA strand, is synthesized continuously toward the replication fork because polymerase can add nucleotides in this direction. This continuously synthesized strand is known as the leading strand. The other strand, complementary to the 5′ to 3′ parental DNA, grows away from the replication fork, so the polymerase must move back toward the replication fork to begin adding bases to a new primer, again in the direction away from the replication fork. It does so until it bumps into the previously synthesized strand and then it moves back again. These steps produce small DNA sequence fragments known as Okazaki fragments, each separated by RNA primer. The strand with the Okazaki fragments is known as the lagging strand, and its synthesis is said to be discontinuous.

The leading strand can be extended from one primer alone, whereas the lagging strand needs a new primer for each of the short Okazaki fragments. The overall direction of the lagging strand will be 3′ to 5′, and that of the leading strand 5′ to 3′. A protein called the sliding clamp holds the DNA polymerase in place as it continues to add nucleotides. The sliding clamp is a ring-shaped protein that binds to the DNA and holds the polymerase in place. Beyond its role in initiation, topoisomerase also prevents the overwinding of the DNA double helix ahead of the replication fork as the DNA is opening up; it does so by causing temporary nicks in the DNA helix and then resealing it. As synthesis proceeds, the RNA primers are replaced by DNA. The primers are removed by the exonuclease activity of DNA polymerase I, and the gaps are filled in. The nicks that remain between the newly synthesized DNA (that replaced the RNA primer) and the previously synthesized DNA are sealed by the enzyme DNA ligase that catalyzes the formation of covalent phosphodiester linkage between the 3′-OH end of one DNA fragment and the 5′ phosphate end of the other fragment, stabilizing the sugar-phosphate backbone of the DNA molecule.

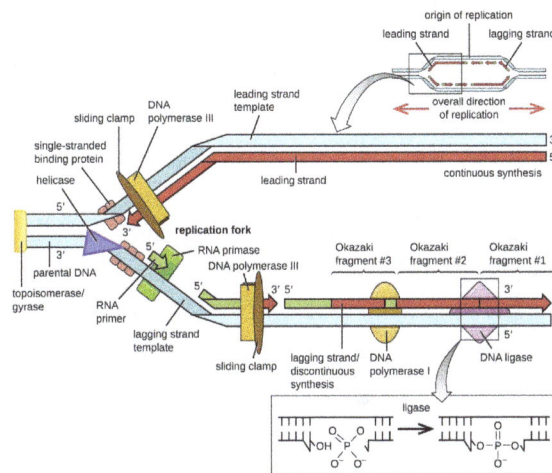

Figure: At the origin of replication, topoisomerase II relaxes the supercoiled chromosome.

Two replication forks are formed by the opening of the double-stranded DNA at the origin, and helicase separates the DNA strands, which are coated by single-stranded binding proteins to keep the strands separated. DNA replication occurs in both directions. An RNA primer complementary to the parental strand is synthesized by RNA primase and is elongated by DNA polymerase III through the addition of nucleotides to the 3′-OH end. On the leading strand, DNA is synthesized continuously, whereas on the lagging strand, DNA is synthesized in short stretches called Okazaki fragments. RNA primers within the lagging strand are removed by the exonuclease activity of DNA polymerase I, and the Okazaki fragments are joined by DNA ligase.

Termination

Once the complete chromosome has been replicated, termination of DNA replication must occur. Although much is known about initiation of replication, less is known about the termination process. Following replication, the resulting complete circular genomes of prokaryotes are concatenated, meaning that the circular DNA chromosomes are interlocked and must be separated from each other. This is accomplished through the activity of bacterial topoisomerase IV, which introduces double-stranded breaks into DNA molecules, allowing them to separate from each other; the enzyme then reseals the circular chromosomes. The resolution of concatemers is an issue unique to prokaryotic DNA replication because of their circular chromosomes. Because both bacterial DNA gyrase and topoisomerase IV are distinct from their eukaryotic counterparts, these enzymes serve as targets for a class of antimicrobial drugs called quinolones.

DNA Replication in Eukaryotes

Eukaryotic genomes are much more complex and larger than prokaryotic genomes and are typically composed of multiple linear chromosomes. The human genome, for example, has 3 billion base pairs per haploid set of chromosomes, and 6 billion base pairs are inserted during replication. There are multiple origins of replication on each eukaryotic chromosome; the human genome has 30,000 to 50,000 origins of replication. The rate of replication is approximately 100 nucleotides per second—10 times slower than prokaryotic replication.

Table: Comparison of Bacterial and Eukaryotic Replication		
Property	Bacteria	Eukaryotes
Genome structure	Single circular chromosome	Multiple linear chromosomes
Number of origins per chromosome	Single	Multiple
Rate of replication	1000 nucleotides per second	100 nucleotides per second
Telomerase	Not present	Present
RNA primer removal	DNA pol I	RNase H
Strand elongation	DNA pol III	pol δ, pol ε

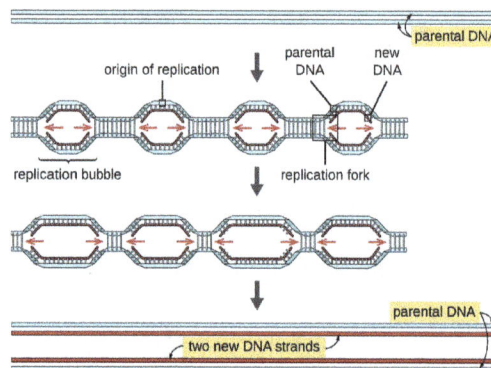

Figure: Eukaryotic chromosomes are typically linear, and each contains multiple origins of replication.

The essential steps of replication in eukaryotes are the same as in prokaryotes. Before replication can start, the DNA has to be made available as a template. Eukaryotic DNA is highly supercoiled and packaged, which is facilitated by many proteins, including histones. At the origin of replication, a prereplication complex composed of several proteins, including helicase, forms and recruits other enzymes involved in the initiation of replication, including topoisomerase to relax supercoiling, single-stranded binding protein, RNA primase, and DNA polymerase. Following initiation of replication, in a process similar to that found in prokaryotes, elongation is facilitated by eukaryotic DNA polymerases. The leading strand is continuously synthesized by the eukaryotic polymerase enzyme pol δ, while the lagging strand is synthesized by pol ε. A sliding clamp protein holds the DNA polymerase in place so that it does not fall off the DNA. The enzyme ribonuclease H (RNase H), instead of a DNA polymerase as in bacteria, removes the RNA primer, which is then replaced with DNA nucleotides. The gaps that remain are sealed by DNA ligase.

Because eukaryotic chromosomes are linear, one might expect that their replication would be more straightforward. As in prokaryotes, the eukaryotic DNA polymerase can add nucleotides only in the 5′ to 3′ direction. In the leading strand, synthesis continues until it reaches either the end of the chromosome or another replication fork progressing in the opposite direction. On the lagging strand, DNA is synthesized in short stretches, each of which is initiated by a separate primer. When the replication fork reaches the end of the linear chromosome, there is no place to make a primer for the DNA fragment to be copied at the end of the chromosome. These ends thus remain unpaired and, over time, they may get progressively shorter as cells continue to divide.

The ends of the linear chromosomes are known as telomeres and consist of noncoding repetitive sequences. The telomeres protect coding sequences from being lost as cells continue to divide. In humans, a six base-pair sequence, TTAGGG, is repeated 100 to 1000 times to form the telomere. The discovery of the enzyme telomerase clarified our understanding of how chromosome ends are maintained. Telomerase contains a catalytic part and a built-in RNA template. It attaches to the end of the chromosome, and complementary bases to the RNA template are added on the 3′ end of the DNA strand. Once the 3′ end of the lagging strand template is sufficiently elongated, DNA polymerase can add the nucleotides complementary to the ends of the chromosomes. In this way, the ends of the chromosomes are replicated. In humans, telomerase is typically active in germ cells and adult stem cells; it is not active in adult somatic cells and may be associated with the aging of these cells. Eukaryotic microbes including fungi and protozoans also produce telomerase to maintain chromosomal integrity.

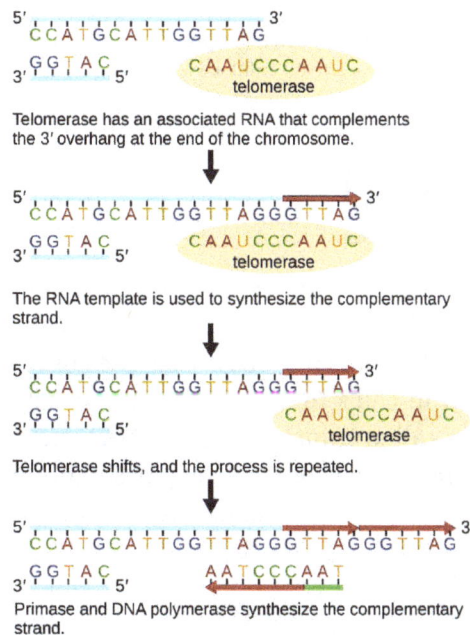

Figure: In eukaryotes, the ends of the linear chromosomes are maintained by the action of the telomerase enzyme.

DNA Replication of Extrachromosomal Elements: Plasmids and Viruses

To copy their nucleic acids, plasmids and viruses frequently use variations on the pattern of DNA replication described for prokaryote genomes. For more information on the wide range of viral replication strategies.

Rolling Circle Replication

Whereas many bacterial plasmids replicate by a process similar to that used to copy the bacterial chromosome, other plasmids, several bacteriophages, and some viruses of eukaryotes use rolling circle replication. The circular nature of plasmids and the circularization of some viral genomes on infection make this possible. Rolling circle replication begins with the enzymatic nicking of one strand of the double-stranded circular molecule at the double-stranded origin (dso) site. In bacteria, DNA polymerase III binds to the 3'-OH group of the nicked strand and begins to unidirectionally replicate the DNA using the un-nicked strand as a template, displacing the nicked strand as it does so. Completion of DNA replication at the site of the original nick results in full displacement of the nicked strand, which may then recircularize into a single-stranded DNA molecule. RNA primase then synthesizes a primer to initiate DNA replication at the single-stranded origin (sso) site of the single-stranded DNA (ssDNA) molecule, resulting in a double-stranded DNA (dsDNA) molecule identical to the other circular DNA molecule.

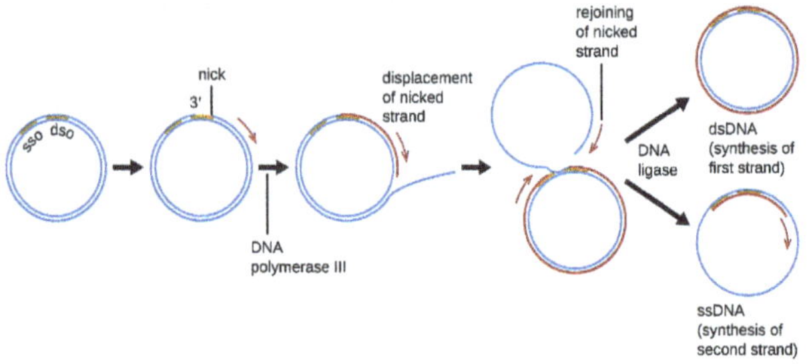

Figure: The process of rolling circle replication results in the synthesis of a single new copy of the circular DNA molecule, as shown here.

Mitochondrial DNA

Mitochondrial DNA contains 37 genes, all of which are essential for normal mitochondrial function. Thirteen of these genes provide instructions for making enzymes involved in oxidative phosphorylation. Oxidative phosphorylation is a process that uses oxygen and simple sugars to create adenosine triphosphate (ATP), the cell's main energy source. The remaining genes provide instructions for making molecules called transfer RNA (tRNA) and ribosomal RNA (rRNA), which are chemical cousins of DNA. These types of RNA help assemble protein building blocks (amino acids) into functioning proteins.

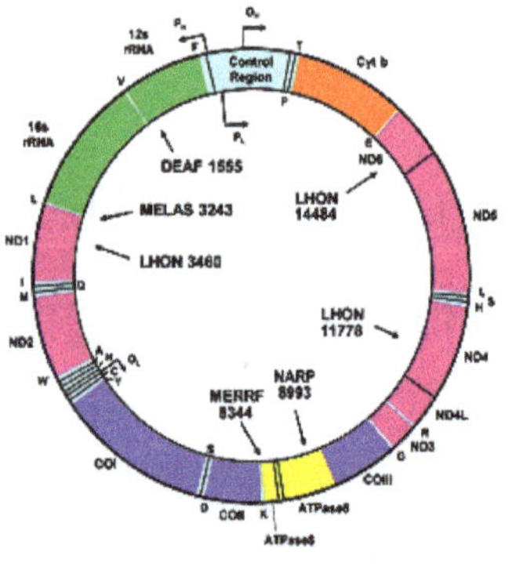

Mitochondrial DNA is typically diagrammed as a circular structure with genes and regulatory regions labeled.

Mitochondrial Inheritance

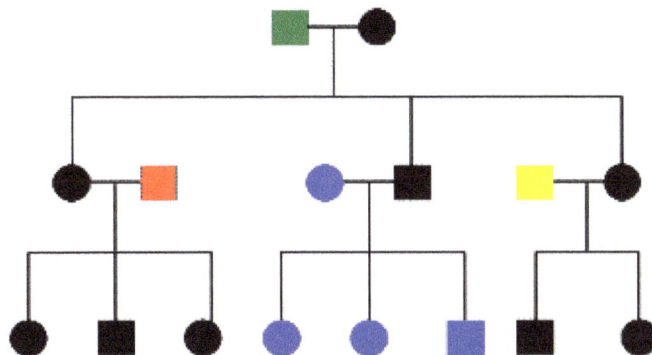

Figure: Inheritance of the maternal mitochondrial genome.
The mitochondrial genome is inherited from the mother in each generation.

Mitochondrial DNA in humans is always inherited from a person's mother. As a result, we share our mitochondrial DNA sequence with our mothers, brothers, sisters, maternal grandmothers, maternal aunts and uncles, and other maternal relatives. Due to the high mutation rates associated with mitochondrial DNA, significant variability exists in mitochondrial DNA sequences among unrelated individuals. However, the mitochondrial DNA sequences of maternally related individuals, such as a grandmother and her grandson or granddaughter, are very similar and can be easily matched.

Mitochondrial DNA sequence data has proved extremely useful in human rights cases, as it is a great a tool for establishing the identity of individuals who have been separated from their families. This approach has been very successful for the following reasons:

- A person's mitochondrial DNA sequence is shared with all of his or her maternal relatives, allowing a genetic match even with few surviving relatives.

- Mitochondrial DNA varies greatly between unrelated families, but it should be nearly identical among closely related individuals.

- A given cell contains many more copies of its mitochondrial DNA than its nuclear DNA, which allows researchers to more easily obtain and analyze mitochondrial DNA samples from deceased relatives.

RNA

RNA is a polymer of ribo-nucleoside-phosphates. Its backbone is comprised of alternating ribose and phosphate groups. Ribose is a five carbon sugar with carbons numbered 1' through 5'. A base is attached to the Y position, generally adenine (A), cytosine (C), guanine (G) or uracil (U). Adenine and guanine are purines, cytosine and uracil are pyrimidine's. A phosphate group is attached to the 3′ position of one ribose and the 5′ position of the next. Most cellular RNA is single stranded, although some viruses have double stranded RNA.

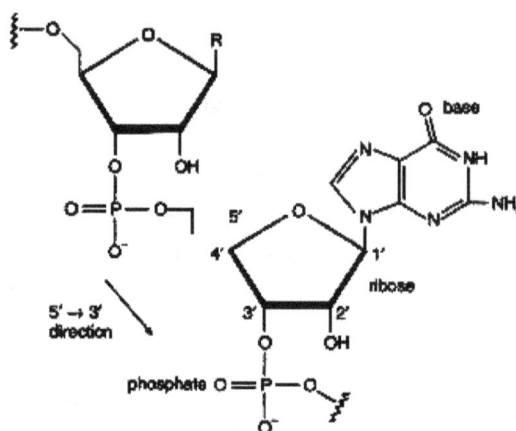

Chemical Structure of RNA

Several other bases are occasionally found in RNAs including: thymine, pseudouridine and methylated cytosine and guanine. However, there are also numerous modified bases and sugars found in RNA that serve many different roles. Pseudouridine (Ψ), in which the linkage between uracil and ribose is changed from a C-N bond to a C-C bond, and ribothymidine (T), are found in various places (most notably in the TΨC loop of tRNA). Thus, it is not technically correct to say that uracil is found in RNA in place of thymine. Another notable modified base is hypoxanthine (a deaminated Guanine base whose nucleoside is called Inosine).

Inosine plays a key role in the Wobble Hypothesis of the Genetic Code. There are nearly 100 other naturally occurring modified nucleosides, of which pseudouridine and nucleosides with 2′-o-methylribose are by far the most common. The specific roles of many of these modifications in RNA are not fully understood. However, it is notable that in ribosomal RNA, many of the post-translational modifications occur in highly functional regions, such as the peptidyl transferase center and the subunit interface, implying that they are important for normal function. The single RNA strand is folded upon itself, either entirely or in certain regions.

In the folded region a majority of the bases are complementary and are joined by hydrogen bonds. This helps in the stability of the molecule. In the unfolded region the bases have no complements. Because of this RNA does not have the purine pyrimidine equality that is found in DNA.

Inside of cells, there are three major types of RNA: messenger RNA (mRNA), transfer RNA (tRNA) and ribosomal RNA (rRNA). There are a number of other types of RNA present in smaller quantities as well, including small nuclear RNA (snRNA), small nucleolar RNA (snoRNA) and the 4.5S signal recognition particle (SRP) RNA. Novel species of RNA continue to be identified. RNA serves a multitude of roles in living cells.

These include: serving as a temporary copy of genes that is used as a template for protein synthesis (mRNA), functioning as adaptor molecules that decode the genetic code (tRNA) and catalyzing the synthesis of proteins (rRNA). There is much evidence implicating RNA structure in biological regulation and catalysis. Interestingly, RNA is the only biological polymer that serves as both a catalyst (like proteins) and as information storage (like DNA).

For this reason, it has be postulated RNA, or an RNA-like molecule, was the basis of life early in evolution. RNA is a nucleic acid polymer consisting of nucleotide monomers that plays several important roles in the processes that translate genetic information from deoxyribonucleic acid (DNA) into protein products; RNA acts as a messenger between DNA and the protein synthesis complexes known as ribosomes, forms vital portions of ribosomes, and acts as an essential carrier molecule for amino acids to be used in protein synthesis.

Synthesis of RNA

There are three types of RNA; messenger RNA (mRNA) or template RNA, ribosomal RNA (rRNA) and soluble RNA (sRNA) or transfer RNA (tRNA). Ribosomal and transfer RNA comprise about 98% of all RNA. All three forms of RNA are made on a DNA template. Transfer RNA and messenger RNA are synthesized on DNA templates of the chromosomes, while ribosomal RNA is derived from nucleolar DNA.

The three types of RNA are synthesized during different stages in early development. Most of the RNA synthesized during cleavage is mRNA. Synthesis of tRNA occurs at the end of cleavage, and rRNA synthesis begins during gastrulation.

Synthesis of RNA is usually catalyzed by an enzyme RNA polymerase using DNA as a template, a process known as transcription. Initiation of transcription begins with the binding of the enzyme to a promoter sequence in the DNA (usually found "upstream" of a gene).

The DNA double helix is unwound by the helicase activity of the enzyme. The enzyme then progresses along the template strand in the 3′ to 5′ direction, synthesizing a complementary RNA molecule with elongation occurring in the 5′ to 3′ direction.

The DNA sequence also dictates where termination of RNA synthesis will occur. RNAs are often modified by enzymes after transcription. For example, a poly (A) tail and a 5′ cap are added to eukaryotic pre-mRNA and introns are removed by the splice some.

There are also a number of RNA-dependent RNA polymerases that use RNA as their template for synthesis of a new strand of RNA. For instance, a number of RNA viruses (such as poliovirus) use this type of enzyme to replicate their genetic material.

Types of RNA

A. Messenger RNA (mRNA)

Messenger RNA (mRNA) is only 5-10% of total RNA present in the cell. This RNA carries genetic information from DNA to the ribosome. Since mRNA is transcribed on DNA (genes), its base sequence is complementary to that of the segment of DNA on which it is transcribed. Usually each gene transcribes its own mRNA. Therefore, there are approximately as many types of mRNA molecules as there are genes.

There may be 1,000 to 10,000 different species of mRNA in a cell. These mRNA types differ only in the sequence of their bases and in length. Messenger RNA is first synthesized by genes as nuclear heterogeneous RNA (hnRNA), being so called because hnRNAs varies enormously in their molecular weight as well as in their nucleotide sequences and lengths, which reflects the different proteins they are destined to code for translation.

The coding sequence of the mRNA determines the amino acid sequence in the protein. In eukaryotic cells, once precursor mRNA (pre-mRNA) has been transcribed from DNA, it is processed to mature mRNA. This removes its introns i.e., non-coding sections of the pre-mRNA. The mRNA is then exported from the nucleus to the cytoplasm, where it is bound to ribosomes and translated into its corresponding protein form with the help of tRNA.

In prokaryotic cells, which do not have nucleus and cytoplasm compartments, mRNA can bind to ribosomes while it is being transcribed from DNA. After a certain period of time the RNA degrades into its component nucleotides with the assistance of ribonucleases. So, mRNA is short lived. When one gene (cistron) codes for a single mRNA strand the mRNA is said to be monocistronic. In many cases, however, several adjacent cistrons may transcribe an mRNA molecule, which is then said to be polycistronic or polygenic (e.g., in prokaryotes). This polycistronic mRNA encode multiple proteins that are separately translated from the same mRNA molecule.

The eukaryotic mRNA is typically monocistronic i.e., only one species of polypeptide chain is translated per mRNA molecule. Most mRNAs contain a significant non-coding segment, that is, a portion that does not direct the assembly of amino acids. For example, approx. 25% of each globin mRNA consists of non-coding, non-translated regions. Non-coding portions are found on both the 5′ and 3′ ends of a messenger RNA and contain sequences that have important regulatory roles.

Messenger RNA is always single stranded. It contains mostly the bases adenine, guanine, cytosine and uracil. There are few unusual substituted bases. Although there is a certain amount of random coiling in extracted mRNA, there is no base pairing. In fact base pairing in the mRNA strand destroys its biological activity. It is heterogeneous in size and sequence. The cap (5′-G) is added to the mRNA after transcription.

The addition of 5′ G is catalyzed by a nuclear enzyme, guanylyl transferase. The cap is linked to the 5′ terminus of the mRNA through an unusual 5′, 5′- triphospahe linkage. The 5′ cap is formed by condensation of a molecule of GTP with the triphosphate at the 5′ end of the transcript. The guanine is subsequently methylated at N-7 to form 7-methylguanosine. Additional methyl groups are added to the 2′ hydroxyls (—OH) of the first and second nucleotides adjacent to the cap. The methyl groups are derived from S-adenosylmethionine. This cap serves to identify this RNA molecule as an mRNA to the translational machinery.

In addition, most mRNA molecules contain a poly-Adenosine tail (poly 'A' tail) at the 3′ end. Both the 5′ cap and the 3′ tail are added after the RNA is transcribed and contribute to the stability of the mRNA in the cell. Therefore, the molecular weight of mRNAs varies from some hundreds to Thousands of Daltons.

The molecular weight of an average sized mRNA molecule is about 500,000, and its sedimentation coefficient is 8S. mRNA varies greatly in length and molecular weight. Since most proteins contain at least a hundred amino acid residues, mRNA must have at least 300 (100X3) nucleotides on the basis of the triplet code. In E. coli the average mRNA strand has 900 to 1,500 nucleotide units which would code polypeptide chains of 300-500 amino acids. Molecules containing 12,000 nucleotide units are also known.

The structure of prokaryotic and eukaryotic mRNA is as follows:

1. 5′ Cap

At the 5′ end of the mRNA molecule in most eukaryote cells and animal virus molecules is found a 'cap'. This cap is formed by the methylation of any of the four nucleotides. The cap helps mRNA to bind to the ribosomes. Without the cap mRNA molecules bind very poorly to the ribosomes. The bacterial mRNA does not have 5'cap. But they have specific ribosome binding site about six nucleotide long, which occurs at several places in the mRNA molecules. These are located at four nucleotides upstream from AUC.

2. Non-coding Regions

As the name indicates these regions do not code for protein. There are two non-coding regions. First non-coding region (NCI) is followed by a 5'cap and is 10 to 100 nucleotides in length. This region is rich in A and U residues. The second non- coding region (NC2) is followed by termination codon and is 50-150 nucleotides long and contains an AAUAAA residues.

3. Initiation Codon

Initiation codon is AUG in both prokaryotes and eukaryotes. Bacterial ribosomes bind directly to the AUG region of the mRNA to start the protein synthesis, whereas this is not there in the case of eukaryotes.

4. The Coding Region

It consists of about 1,500 nucleotides. This region is responsible for coding protein with several ribosomes. The combination of mRNA strand with several ribosomes is called polyribosomes.

5. The Termination Codon

The termination codon is required to give the signal to stop protein synthesis.

6. The poly (A) Sequence

The non-coding region II is followed by poly (A) sequence in the eukaryotic mRNA. The prokaryotic mRNAs lack poly (A). The poly (A) sequences of 200-250 nucleotides are present at 3'OH end of mRNA. Poly (A) sequences are added when mRNA is present inside the nucleus. The function of poly (A) sequence in translation is unknown.

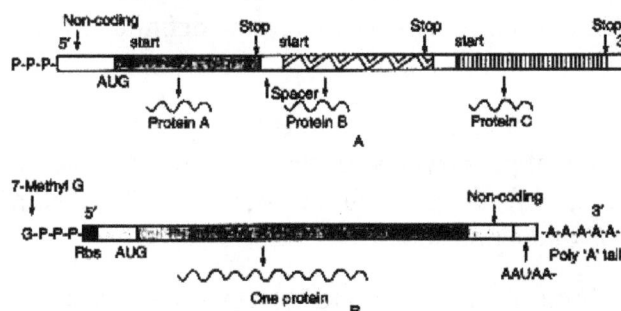

structure of prokaryolic (A) and eukaryotic (B) mRNA showing gene product. this requires processing of mRNA in eukaryotic before translation.

Difference between Prokaryotic and Eukaryotic mRNA

Prokaryotic mRNA

1. Translation begins when the mRNA is still being transcribed on DNA.

2. Prokaryote mRNA are very short lived. It constantly under goes breakdown to its constituent ribonucleotides by ribonucleases.

3. In Prokaryotic mRNA are polycistronic.

4. The mRNA undergo very little processing after being transcribed.

5. Prokaryotic mRNA do not have poly (A) tail.

Eukaryotic mRNA

1. Translation begins when the transcription is Completed.

2. Eukaryotic nRNAs are long lived. Thus are metabolically stable.

3. In Eukaryotic mRna are monocistronic.

4. The mRNA undergoes several processing after being transcribed such as polyadenylation capping and methylation.

5. Eukaryotic mRNA have poly (A) tail.

B. Ribosomal RNA (rRNA)

Ribosomal RNA is extremely abundant and makes up to 80% of RNA found in a typical eukaryotic cytoplasm. Ribosomal RNA consists of a single strand twisted upon itself in some regions. It has helical regions connected by intervening single strand regions. The helical regions may show presence or absence of positive interaction. In the helical region most of the base pairs are complementary, and are joined by hydrogen bonds. In the unfolded single strand regions the bases have no complements.

In the cytoplasm, ribosomal RNA and protein combine to form a nucleoprotein called a ribosome. Ribosomal RNA (rRNA) is a component of the ribosomes. The base sequence of rRNA is complementary to that of the region of DNA where it is synthesized. Ribosomal RNA is formed from only a small section of the DNA molecule, and hence there is no definite base relationship between rRNA and DNA as a whole.

Primarily there are two types of ribosomes, one is 70s (prokaryotes) and the other is 80s (eukaryotes). The 70S ribosome of prokaryotes consists of a 30S subunit and a 50S subunit. The 30S subunit contains 16S rRNA, while the 50S subunit contains 23S and 5S rRNA. The 80S eukaryote ribosome consists of a 40S and a 60S subunit.

In vertebrates the 40S subunit contains 18S rRNA, while the 60S subunit contains 28-29S, 5.8S and 5S rRNA. In plants and invertebrates, the 40S subunit contains 16-18S RNA, while the 60S subunit contains 25S and 58 and 5.8S rRNA.

C. Transfer or tRNA

The tRNA molecules are key to the translation process of the mRNA sequence into the amino acid sequence of proteins (at least one type of tRNA for every amino acid). To be precise, the amino-acyl-tRNA-synthase proteins are the 'true' translators of the genetic code into an amino acid sequence. These synthetases acetylate tRNA molecules with the proper amino acid that corresponds to the anti-codon in the structure of the tRNA molecule.

The anti-codon later recognizes the codon, the triple base sequence which 'codes' for the amino acid along the mRNA strand. A failure of properly acetylating the tRNA with the right amino acid results in an amino acid mutation even though the DNA sequence has not been changed. tRNA molecules are small nucleic acids of 60- 95 nucleotides, mostly 76, with a molecular weight 18-20kD, with the secondary structure resembling a clover leaf. Here are a few common features shared by all tRNA molecules found in various organisms.

1. 5′ terminus is always phosphorylated.

2. 7 bp stem, may have non-Watson & Crick pairing (like GU) acceptor or amino acid stem at 3′ terminus in which last three nucleotides are CCA-3′-OH. These are added after transcription and amino acylation occurs at 3′-OH group of last base 'A'.

3. 3-4 bp stem and loop contains the base dihydrouridine (D) [D- arm].

4. 5 bp stem and loop containing anti-codon triplet [anti-codon arm].

5. 5 bp stem and loop contains sequence T C, standing for [T- arm] pseudouridine.

6. Variable arm (between anti-codon and T-arm) length, 3-21 nucleotides.

7. Contains numerous modified bases (up to 25%) which are all post-transcriptionally modified.

The three dimensional structure of tRNA resembles an L-shaped molecule with the D-arm and anti-codon loop building one stretch and the T-arm and acceptor stem building the other stretch being deposed by ~90 to one another (interstem angle of 82 by X-ray refinement and 92 in an electron microscopy study). The molecule is about 6 nm in each direction with the anti-codon to acceptor 3′-term ends being 7.6 nm apart. The diameter of both arms is about 2.0 to 2.5 nm.

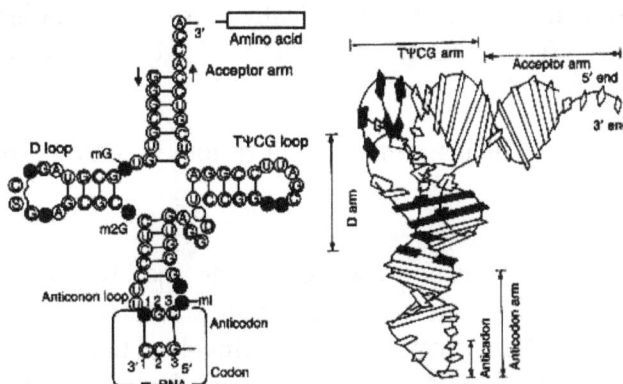

Figure: A 3D structure of tRNA; clover leaf model

Difference between the three types of RNAs:

	Ribosmal RNA (rRNA)	massenger rna (mRNA)	transfer RNA(tRNA)
Percentage of total RNA of cell	-80%	5 to 10%	10-15%
Sedimentation coefficeint	28s. 18s. 5.8s. 5s. 23s. 16s. 5s.	8s.	3.8s
Number of nucleotides	5s RNA: 120 nucleotides. 16-18s RNA: 1.600 to 2.500 nucleotides. 23-28s RNA: 3200 to 5,500 nucleotides	*E. coli:* 900 to 1.500 nucleotides	73 to 93 nucleotides
Unsual bases	Small amount of methylated bases (*E. coli:* 1 per 100-150 nucleotides).	small amount of unsual bases	high content of unusual bases. (*E. coli:* 1 per 3040 nucleotides).
Site of synthesis	Derived from nucleor DNA.	Synthesized in nucleus on DNA template.	Synthesized in nucleus on DNA template.
Begining of synthesis	Synthesis begins at gastrulation and increases as development proceeds.	Some mRNA are found in the ovum. New mRNA is synthesized during early cleavage.	tRNA synthesis occurs at the end of cleavage stages.
Base relationship with DNA	No obvious base relationship to DNA. rRNA is formed from only small section of DNA.	mRNA shows base relationship with DNA. it formed from only small section of DNA.	rRNA also shows base relationship with DNA. it is formed from all sections of DNA.
Function	Unpaired bases may bind mRNA and tRNA to ribosomes	Conveys genetic information froDNA to the ribosomes, where it takes part in protein synthesis.	Adapter for attaching amino acids to mRNA template.

The structural complexity of tRNA is reminiscent of that of a protein with 71 out of 76 bases participating in stacking interaction (of which 42 in double helical stem structures). 9 bp interactions are cross linking the tertiary structure, i.e., they interact with bases from a different stem and loop region. All of these 9 bp are non-Watson-Crick associations and are highly conserved which makes it likely to predict similar structures for all tRNA molecules (in fact, only few tRNA molecules have been crystallized and their structure determined).

New Types of RNA

Besides mRNA, tRNA and rRNA, the three classically known RNA molecules, new types of RNA molecules are discovered in recent years, which are involved in various activities directly or they modulate the activity.

Some of the new types of RNA which are involved in various activities directly are as follows:

1. Non-Coding RNA

RNA genes (sometimes referred to as non-coding RNA or small RNA) are genes that encode RNA that is not translated into a protein. The most prominent examples of RNA genes are transfer RNA (tRNA) and ribosomal RNA (rRNA), both of which are involved in the process of translation. However, since the late 1990s, many new RNA genes have been found, and thus RNA genes may play a much more significant role than previously thought.

In the late 1990s and early 2000, there has been persistent evidence of more complex transcription occurring in mammalian cells (and possibly others). This could point towards a more widespread use of RNA in biology, particularly in gene regulation. A particular class of non-coding RNA, micro RNA, has been found in many metazoans (from Caenorhabditis elegans to Homo sapiens) and clearly plays an important role in regulating other genes.

2. Small Nuclear RNA (snRNA)

About a dozen genes for snRNAs have been described, each present in multiple copies in the genome. Small nuclear RNAs combine with certain U-proteins to form snRNPs. These executive molecules have roles in editing other classes of RNA. The "U" designation was given to the snRNAs because they were found to be rich in uridylic acid.

An important example is the small ribonuclear proteins (snRNPs) that are components of the spliceosomes. Spliceosomes edits introns nitrogen base sequences out of pm RNA forming mRNA. U6 is transcribed in the nucleoplasm by RNA polymerase III, while U1, U2, U4 and U5 are transcribed by RNA polymerase II. The snRNPs are involved in processing of pmRNA.

3. Small Nucleolar RNA (snoRNA)

In eukaryotic cells, rRNA and snRNA are extensively modified and processed in the nucleolus. Much of this activity is affected by snoRNAs. It appears that the coding for snoRNAs lies in introns and other intergenic regions (non-protein coding). Small nucleolar RNA can be considered to be a subgroup of the snRNAs but should not be confused with the snRNAs mediating mRNA splicing, i.e. the spliceosomal RNAs.

snoRNAs have not been found in any prokaryotes. Many snoRNAs and snRNAs have been found to be encoded in introns. It has been suggested that snoRNAs are possibly processed from the out-spliced introns by exonucleases.

4. Short Interfering RNAs (siRNA)

Short interfering RNAs and miRNAs were discovered in different works, but their biogenesis and assembly into RNA-protein complexes and their function in down regulating gene expression are closely related. Short interfering RNAs and mi RNAs share common RNAse III processing enzyme, the dicer enzymes and closely related effector complexes for post-transcriptional repression of protein synthesis. On the other hand, siRNAs and miRNAs differ in their molecular origins. In the cytoplasm the dicer enzymes split the dsRNA primer molecules. The finished siRNAs in animals are usually 21-22 nitrogen bases long, similar to miRNAs.

5. Micro-RNAs (miRNAs)

Micro-RNAs are a class of small, non-coding RNAs that regulate gene expression in a sequence specific manner as required in embryonic development. Micro-RNAs have been found throughout diverse eukaryotes genomes including plants. They can inhibit protein expression by shutting off translation or by targeting mRNA for degradation.

Micro-RNAs genes produce short (~22 nitrogen bases) ss segments that fold over on themselves forming a short section of dsRNA in hairpin like structure. Humans express over 460 genetically encoded miRNAs. These miRNAs make up more than 1 % of human genome and may regulate over 30% of all protein coding genes. Micro-RNAs can pair exactly with a mRNA and cause its cleavage and destruction, or it can pair partially with mRNA and produce translational inhibition (block the ribosome). It is presumed that they are involved in regulating development by controlling as transcriptional factor.

6. Ribozymes

Ribozymes are catalytic RNA enzymes that act to alter covalent structure in other classes of RNAs and certain molecules. They occur in ribosomes, nucleus and chloroplasts of eukaryotic organisms. Some viruses including several bacteriophages also have ribozymes. An optimum concentration of metal ions such as Mg^{++} and K^+ is associated with their effective functioning. Ribozymes generally act as molecular scissors cutting precursor RNA molecules at specific sites. Surprisingly, they also serve as molecular staplers, which ligate or join two RNA molecules together.

Ribozymes are involved in the transformation of large precursor molecules of tRNA, rRNA and mRNA into smaller final products. In their active form, ribozymes are complexed with protein molecules, e.g., the enzyme ribonuclease-P (RNAse-P) is found in all living organisms. Ribonuclease-P is a heterodimer containing one molecule of protein and one molecule of RNA.

This ribozyme cleaves the head 5′ end of the precursors to the tRNAs. These enzymes are involved in autocatalytic splicing of preRNA making contiguous RNA. This cleavage is carried out by RNA part of the heterodimer.

Synthetic ribozymes can be used in genetic engineering applications to stop the expression of any gene in a sequence-specific manner and therefore may be useful in cancer therapy and HIV treatment. Ribozymes can be inserted into the cells has synthetic oligonucleotides with a gene gun or the cell can synthesize them itself with engineered genes.

7. XISTRNA

XIST stands for X-inactive-specific transcript. It is a large (~ 17kb) RNA coded by a gene on the X chromosome (8 exons are involved with human xist). xistRNA accumulates in female somatic cells along the X chromosome containing the active xist gene and proceeds to inactivate nearly all the 100s of genes on that X chromosome. The xistRNA does not travel over to any other chromosome in nucleus. The barr body seen in the cell nucleus with light microscopy is inactive X chromosome covered with xistRNA. The entire xistRNA X-blocking mechanism is very complex.

8. Double-Stranded RNA (dsRNA)

Double-stranded RNA (dsRNA) is RNA with two complementary strands, similar to the DNA found in all 'higher' cells. dsRNA forms the genetic material of some viruses. In eukaryotes, it acts as a trigger to initiate the process of RNA interference and is present as an intermediate step in the formation of siRNAs (small interfering RNAs). siRNAs are often confused with miRNAs; siRNAs are double-stranded, whereas miRNAs are single-stranded.

Although initially single stranded, there are regions of intra-molecular association causing hairpin structures in pre-miRNAs. Very recently, dsRNA has been found to induce gene expression at transcriptional level, a phenomenon named "small RNA induced gene activation (RNAa)". Such dsRNA is called "small activating RNA (saRNA)".

9. RNA Secondary Structures

The functional form of single stranded RNA molecules (like proteins) frequently requires a specific tertiary structure. The scaffold for this structure is provided by secondary structural elements which are hydrogen bonds within the molecule. This leads to several recognizable "domains" of secondary structure like hairpin loops, bulges and internal loops.

The secondary structure of RNA molecules can be predicted computationally by calculating the minimum free energies (MFE) structure for all different combinations of hydrogen bonding's and domains.

References

- Nucleic-acid, science: britannica.com, Retrieved January 7, 2019

- Dna: livescience.com, Retrieved February 3, 2019

- Dna-replication, suny-microbiology: lumenlearning.com, Retrieved July 7, 2019

- Mitochondrial-dna: nih.gov, Retrieved July 18, 2019

- Mtdna-and-mitochondrial-diseases, topicpage, scitable: nature.com, Retrieved August 8, 2019

- Rna-introduction-synthesis-and-types, rna: biologydiscussion.com, Retrieved February 8, 2019

Chapter 6

Enzymes: A Comprehensive Study

The macromolecular biological catalysts that accelerate chemical reactions are known as enzymes. Substrates are the molecules upon which enzymes act and convert it into another molecules known as products. The studies of the chemical reactions that are catalysed by enzymes fall under the domain of enzyme kinetics. This chapter has been carefully written to provide an easy understanding of the various types of enzymes.

Enzymes are proteins that act as catalysts within living cells. Catalysts increase the rate at which chemical reactions occur without being consumed or permanently altered themselves. A chemical reaction is a process that converts one or more substances (known as reagents, reactants, or substrates) to another type of substance (the product). As a catalyst, an enzyme can facilitate the same chemical reaction over and over again.

Structure and Function

Like all proteins, enzymes are composed of one or more long chains of interconnected amino acids. Each enzyme possesses a unique sequence of amino acids that causes it to fold into a characteristic shape. An enzyme's amino acid sequence is determined by a specific gene in the cell's nucleus. This ensures that each copy of the enzyme is the same as all others.

On the surface of each enzyme is a special cleft called the active site, which provides a place where reagents can 'meet' and interact. Much like a lock and its key, an enzyme's active site will only accommodate certain reagents, and only one type of chemical reaction can be catalyzed by a given enzyme.

For example, during the manufacture of hemoglobin (the oxygen-carrying pigment in your red blood cells), a single atom of iron must be inserted into the center of the molecule to make it functional. An enzyme called ferrochelatase brings the reagents (iron and the empty molecule) together, catalyzes their union, and releases an iron-containing molecule. This is the only reaction catalyzed by ferrochelatase. Keep in mind that enzymes can combine reagents (as in the synthesis of hemoglobin), they can split a single reagent into multiple products, or they can simply transform a single reagent into a single product that looks different from the original reagent.

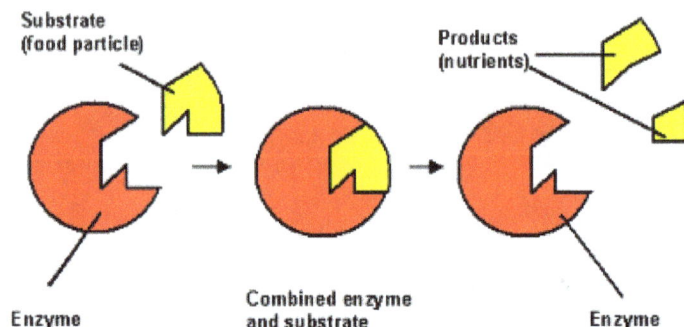

Substrate (food particle)

Products (nutrients)

Enzyme

Combined enzyme and substrate

Enzyme

When reagents enter an enzyme's active site, the enzyme undergoes a temporary change in shape that encourages interaction between the reagents. Upon completion of the chemical reaction, a specific product is released from the active site, the enzyme resumes its original conformation, and the reaction can begin again with new reagents.

Many enzymes are incorporated into metabolic pathways. A metabolic pathway is a series of chemical reactions that transform one or more reagents into an end-product that's needed by the cell. The enzymes in a metabolic pathway--much like people passing a pail of water along a bucket brigade--move a reagent along until the end-product is produced. A metabolic pathway can be quite short, or it can have many steps and multiple enzymes. The metabolic pathway that converts tryptophan (an amino acid found in dietary protein) to serotonin (a chemical that's necessary for normal brain function) is only two steps long.

In order to function, many enzymes require the help of cofactors or coenzymes. Cofactors are often metal ions, such as zinc, copper, iron, or magnesium. Magnesium, one of the most common cofactors, activates hundreds of enzymes, including those that manufacture DNA and many that help metabolize carbohydrates.

Many coenzymes are derived from vitamins. In fact, one of the main reasons you need vitamins in your diet is to supply the raw material for essential coenzymes. For example, vitamin C is needed by the enzyme that produces collagen and builds healthy skin, a coenzyme derived from vitamin B12 is necessary for synthesizing the insulation around your nerve cells, and a vitamin B6-based coenzyme is vital for producing serotonin.

Coenzymes and cofactors bind to the active sites of enzymes, and they participate in catalysis, but they are not generally considered reagents, nor do they become part of the product(s) of the reaction. In many cases, cofactors and coenzymes function as intermediate carriers of electrons, specific atoms, or functional groups that are transferred during the overall reaction.

Types of Enzymes

Oxidoreductases

Oxidoreductases are a class of enzymes that catalyze oxidoreduction reactions. Oxidoreductases catalyze the transfer of electrons from one molecule (the oxidant) to another molecule (the reductant). Oxidoreductases catalyze reactions similar to the following, $A^- + B \rightarrow A + B^-$ where A is the oxidant and B is the reductant. Oxidoreuctases can be oxidases or dehydrogenases. Oxidases are enzymes involved when molecular oxygen acts as an acceptor of hydrogen or electrons. Whereas, dehydrogenases are enzymes that oxidize a substrate by transferring hydrogen to an acceptor that is either $NAD^+/NADP^+$ or a flavin enzyme. Other oxidoreductases include peroxidases, hydroxylases, oxygenases, and reductases. Peroxidases are localized in peroxisomes, and catalyzes the reduction of hydrogen peroxide. Hydroxylases add hydroxyl groups to its substrates. Oxygenases incorporate oxygen from molecular oxygen into organic substrates. Reductases catalyze reductions, in most cases reductases can act like an oxidases.

Oxidoreductase enzymes play an important role in both aerobic and anaerobic metabolism. They can be found in glycolysis, TCA cycle, oxidative phosphorylation, and in amino acid metabolism. In glycolysis, the enzyme glyceraldehydes-3-phosphate dehydrogenase catalyzes the reduction of

NAD$^+$ to NADH. In order to maintain the re-dox state of the cell, this NADH must be re-oxidized to NAD$^+$, which occurs in the oxidative phosphorylation pathway. Additional NADH molecules are generated in the TCA cycle. The product of glycolysis, pyruvate enters the TCA cycle in the form of acetyl-CoA. During anaerobic glycolysis, the oxidation of NADH occurs through the reduction of pyruvate to lactate. The lactate is then oxidized to pyruvate in muscle and liver cells, and the pyruvate is further oxidized in the TCA cycle. All twenty of the amino acids, except leucine and lysine, can be degraded to TCA cycle intermediates. This allows the carbon skeletons of the amino acids to be converted into oxaloacetate and subsequently into pyruvate. The gluconeogenic pathway can then utilize the pyruvate formed.

Glycoside Hydrolases

Glycoside hydrolases are enzymes that catalyze the hydrolysis of the glycosidic linkage of glycosides, leading to the formation of a sugar hemiacetal or hemiketal and the corresponding free aglycon. Glycoside hydrolases are also referred to as glycosidases, and sometimes also as glycosyl hydrolases. Glycoside hydrolases can catalyze the hydrolysis of O$^-$, N$^-$ and S-linked glycosides.

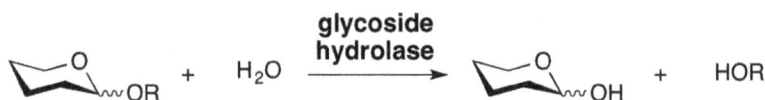

Classification

Glycoside hydrolases can be classified in many different ways:

Endo/exo

exo- and endo- refers to the ability of a glycoside hydrolase to cleave a substrate at the end (most frequently, but not always the non-reducing end) or within the middle of a chain. For example, most cellulases are endo-acting, whereas LacZ β-galactosidase from E. coli is exo-acting. A general sub-site nomenclature exists to demarcate substrate binding in glycosidase active-sites.

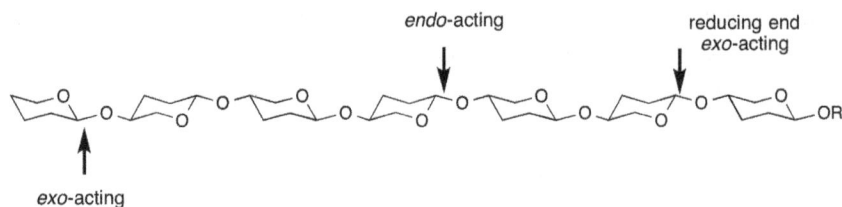

Enzyme Commission (EC) number

EC numbers are codes representing the Enzyme Commission number. This is a numerical classification scheme for enzymes, based on the chemical reactions they catalyze. Every EC number is associated with a recommended name for the respective enzyme. EC numbers do not specify enzymes, but enzyme-catalyzed reactions. If different enzymes (for instance from different organisms) catalyze the same reaction, then they receive the same EC number. A necessary consequence of the EC classification scheme is that codes can be applied only to enzymes for which a function has been biochemically identified. Additionally, certain enzymes can catalyze reactions that fall in more than one class. These enzymes must bear more than one EC number.

Mechanistic Classification

Two reaction mechanisms are most commonly found for the retaining and inverting enzymes, as

Retaining glycoside hydrolases:

Inverting glycoside hydrolases:

Sequence-based Classification

Sequence-based classification uses algorithmic methods to assign sequences to various families. The glycoside hydrolases have been classified into more than 100 families ; this is permanently available through the Carbohydrate Active enZyme database. Each family (GH family) contains proteins that are related by sequence, and by corollary, fold. This allows a number of useful predictions to be made since it has long been noted that the catalytic machinery and molecular mechanism is conserved for the vast majority of the glycosidase families as well as the geometry around the glycosidic bond (irrespective of naming conventions). Usually, the mechanism used (ie retaining or inverting) is conserved within a GH family. One notable exception is the glycoside hydrolases of family GH97, which contains both retaining and inverting enzymes; a glutamate acts as a general base in inverting members, whereas an aspartate likely acts as a catalytic nucleophile in retaining members. Another mechanistic curiosity are the glycoside hydrolases of familes GH4 and GH109 which operate through an NAD-dependent hydrolysis mechanism that proceeds through oxidation-elimination-addition-reduction steps via anionic transition states. This allows a single enzyme to hydrolyze both α- and β-glycosides.

Classification of families into larger groups, termed 'clans' has been proposed. A 'clan' is a group of families that possess significant similarity in their tertiary structure, catalytic residues and mechanism. Families within clans are thought to have a common evolutionary ancestry. For an updated table of glycoside hydrolase clans see the CAZy Database.

Mechanism

Two reaction mechanisms are most commonly found for the retaining and inverting enzymes, as

However several interesting variations on these mechanisms have been found, and one fundamentally different mechanism, catalyzed by an NADH cofactor, has been discovered in recent years.

Inverting Glycoside Hydrolases

Hydrolysis of a glycoside with net inversion of anomeric configuration is generally achieved via a one step, single-displacement mechanism involving oxocarbenium ion-like transition states. The reaction typically occurs with general acid and general base assistance from two amino acid side chains, normally glutamic or aspartic acids, that are typically located 6-11 A apart.

Inverting mechanism for an α-glycosidase:

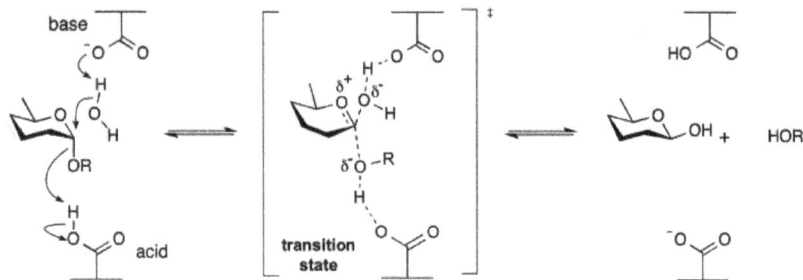

Inverting mechanism for a β-glycosidase:

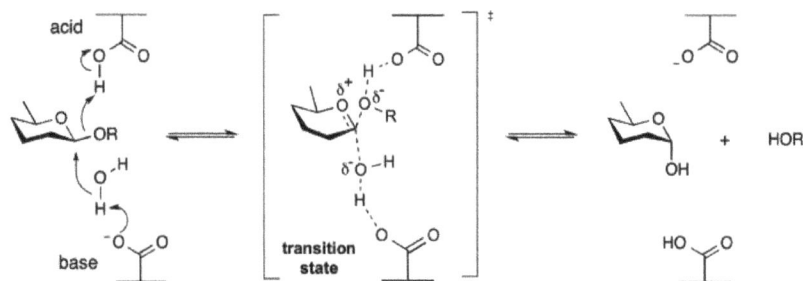

Glycosyl-phosphate Cleaving Enzymes that Lack a General Acid

A subset of family GH92 α-mannosidases catalyze the hydrolysis of mannose-1-phosphate linkages found in the mannose-1-phosphate-6-mannose groups of yeast mannoproteins. In these enzymes the usual general acid glutamic acid found in other members of this family is replaced by a glutamine. It has been suggested that the phosphate aglycon is a sufficiently good leaving group to be able to cleave in the first glycosylation step to form the glycosyl enzyme intermediate without the requirement of an acid catalyst. This replacement may also reduce charge repulsion between the glutamic acid residue and the anionic phosphate aglycon. A related example may be found in the case of the family GH1 myrosinases.

Retaining Glycoside Hydrolases

Classical Koshland Retaining Mechanism

Hydrolysis with net retention of configuration is most commonly achieved via a two step, double-displacement mechanism involving a covalent glycosyl-enzyme intermediate. Each step passes through an oxocarbenium ion-like transition state. Reaction occurs with acid/base and nucleophilic assistance provided by two amino acid side chains, typically glutamate or aspartate, located 5.5 A apart. In the first step (often called the glycosylation step), one residue plays the role of a nucleophile, attacking the anomeric centre to displace the aglycon and form a glycosyl enzyme intermediate. At the same time the other residue functions as an acid catalyst and protonates the glycosidic oxygen as the bond cleaves. In the second step (known as the deglycosylation step), the glycosyl enzyme is hydrolyzed by water, with the other residue now acting as a base catalyst deprotonating the water molecule as it attacks. The pKa value of the acid/base group cycles between high and low values during catalysis to optimize it for its role at each step of catalysis. In the case of sialidases, the catalytic nucleophile is a tyrosine residue. This mechanism was originally proposed by Dan Koshland, although at the time the identities of the residues was unclear.

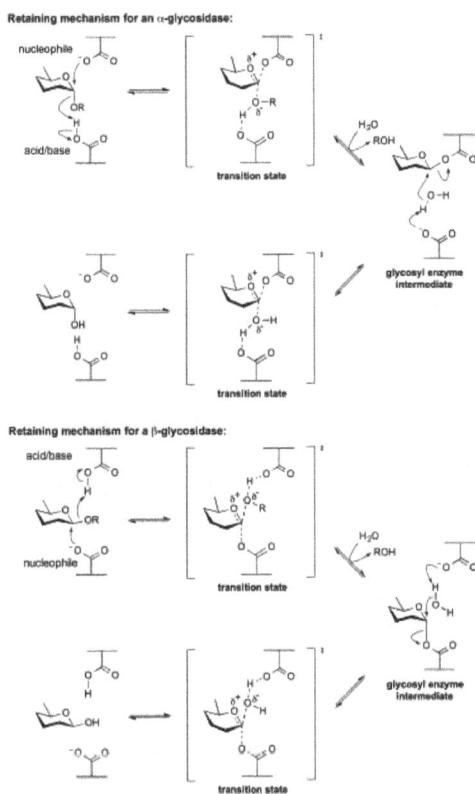

Neighboring Group Participation

Enzymes of glycoside hydrolase families 18, 20, 25, 56, 84, and 85 hydrolyse substrates containing an N-acetyl (acetamido) or N-glycolyl group at the 2-position. These enzymes have no catalytic nucleophile: rather they utilize a mechanism in which the 2-acetamido group acts as an intramolecular nucleophile. Neighboring group participation by the 2-acetamido group leads to formation of an oxazoline (or more strictly an oxazolinium ion) intermediate. This mechanism was deduced from X-ray structures of complexes of chitinases with natural inhibitors, from the potent inhibition afforded by a stable thiazoline analogue of the oxazoline and from detailed mechanistic analyses using substrates of modified reactivity. Typically, a stabilizing residue (a carboxylate) stabilizes the charge development in the transition state. Not all enzymes that cleave substrates possessing a 2-acetamido group utilize a neighboring groups participation mechanism; enzyme of glycoside hydrolase families 3 and 22 utilize a classical retaining mechanism with an enzymic nucleophile. Other hexosaminidases such as those of Glycoside Hydrolase Family 19 utilize an inverting mechanism.

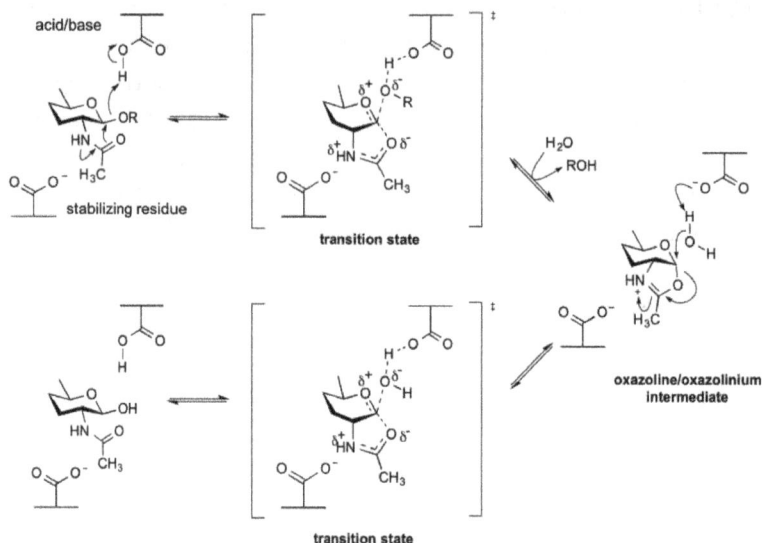

Myrosinases: Retaining Glycoside Hydrolases that Lack a General Acid and Utilize an Exogenous Base

Glycoside hydrolases termed myrosinases catalyze the hydrolysis of anionic thioglycosides (gluco-sinolates) found in plants. They are found in Glycoside Hydrolase Family 1. In these enzymes the usual acid/base glutamic acid found in other members of this family is replaced by a glutamine. This likely reduces charge repulsion between the anionic aglycon sulfate. The unusual aglycon is a sufficiently good leaving group to be able to cleave in the first glycosylation step to form the glycosyl enzyme intermediate without the requirement of an acid catalyst. However, since a base catalyst is required for the second step (hydrolysis or deglycosylation) these enzymes require an alternative basic group. This is provided by the co-enzyme L-ascorbate.

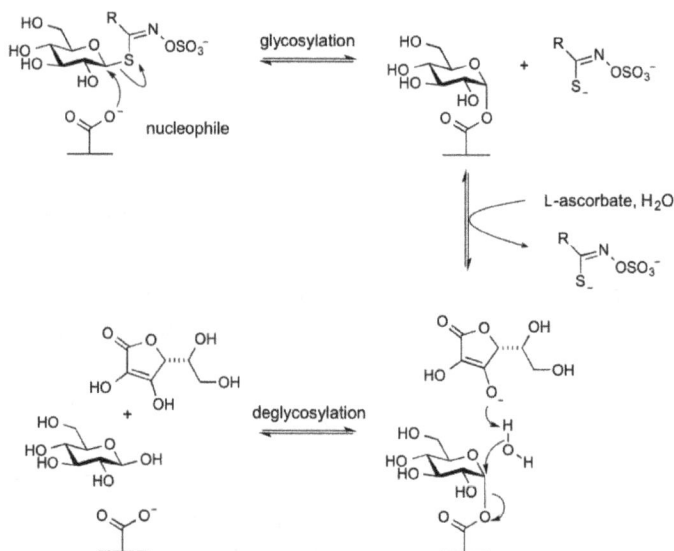

Alternative Nucleophiles

Several groups of retaining glycosidases use atypical nucleophiles. These include the sialidases and trans-sialidases of glycoside hydrolase families 33 and 34, and 2-keto-3-deoxy-d-lyxo-heptulosaric acid hydrolases of 143. Glycoside hydrolases of these families utilize a tyrosine as a catalytic nucleophile, which is believed to be activated by an adjacent base residue. A rationale for this unusual difference is that the use of a negatively charged carboxylate as a nucelophile will be disfavoured as the anomeric centre is itself negatively charged, and thus charge repulsion interferes. A tyrosine residue is a neutral nucleophile, but requires a general base to enhance its nucleophilicity. This mechanism was implied from X-ray structures, and was supported by experiments involving trapping of the intermediate with fluorosugars followed by peptide mapping and then crystallography, as well as via mechanistic studies on mutants.

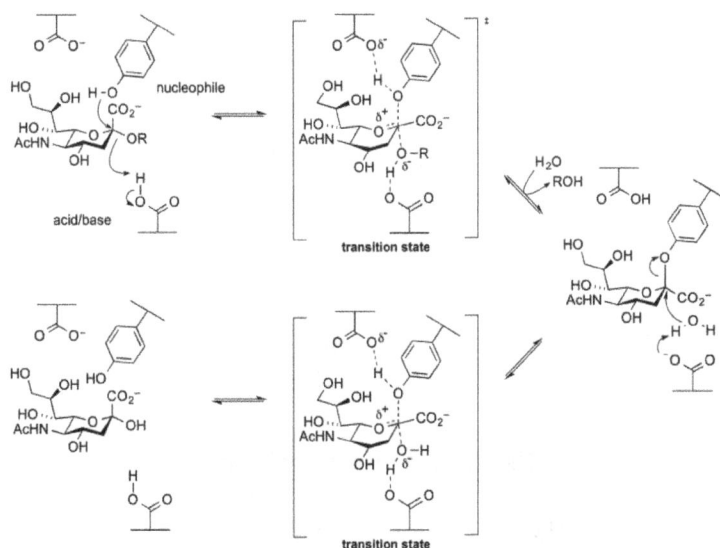

NAD-dependent Hydrolysis

The glycoside hydrolases of family 4 and 109 use a mechanism that requires an NAD cofactor, which remains tightly bound throughout catalysis. The mechanism proceeds via anionic transition states with elimination and redox steps rather than the classical mechanisms proceeding throughoxocarbenium ion-like transition states. for a 6-phospho-β-glucosidase, the mechanism involves an initial oxidation of the 3-hydroxyl of the substrate by the enzyme-bound NAD cofactor. This increases the acidity of the C2 proton such that an E1cb elimination can occur with assistance from an enzymatic base. The α,β-unsaturated intermediate formed then undergoes addition of water at the anomeric centre and finally the ketone at C3 is reduced to generate the free sugar product. Thus, even though glycosidic bond cleavage occurred via an elimination mechanism, the overall reaction is hydrolysis. This mechanism was elucidated through a combination of stereochemical studies by NMR, kinetic isotope effects, linear free energy relationships, X-ray crystallography and UV/Vis spectrophotometry.

Isomerase

Isomerases are a general class of enzymes that convert a molecule from one isomer to another.

Isomerases facilitate intramolecular rearrangements in which bonds are broken and formed. The general form of such a reaction is as follows:

A–B → B–A

There is only one substrate yielding one product. This product has the same molecular formula as the substrate but differs in bond connectivity or spatial arrangement. Isomerases catalyze reactions across many biological processes, such as in glycolysis and carbohydrate metabolism.

Isomerization

Examples of Isomers

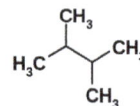

(1)

(2)   (3)

(4)   (5)

The structural isomers of hexane

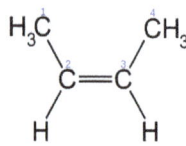 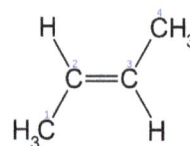

cis-but-2-ene   trans-but-2-ene

Cis-2-butene and Trans-2-butene

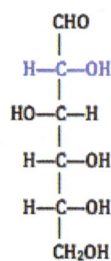 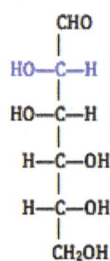

D-Glucose   D-Mannose

Epimers: D-glucose and D-mannose

Isomerases catalyze changes within one molecule. They convert one isomer to another, meaning that the end product has the same molecular formula but a different physical structure. Isomers themselves exist in many varieties but can generally be classified as structural isomers or stereo-isomers. Structural isomers have a different ordering of bonds and/or different bond connectivity from one another, as in the case of hexane and its four other isomeric forms (2-methylpentane, 3-methylpentane, 2,2-dimethylbutane, and 2,3-dimethylbutane).

Stereoisomers have the same ordering of individual bonds and the same connectivity but the three-dimensional arrangement of bonded atoms differ. For example, 2-butene exists in two isomeric forms: *cis*-2-butene and *trans*-2-butene. The sub-categories of isomerases containing racemases, epimerases and cis-trans isomers are examples of enzymes catalyzing the interconversion of stereoisomers. Intramolecular lyases, oxidoreductases and transferases catalyze the interconversion of structural isomers.

The prevalence of each isomer in nature depends in part on the isomerization energy, the difference in energy between isomers. Isomers close in energy can interconvert easily and are often seen in comparable proportions. The isomerization energy, for example, for converting from a stable *cis* isomer to the less stable *trans* isomer is greater than for the reverse reaction, explaining why in the absence of isomerases or an outside energy source such as ultraviolet radiation a given *cis* isomer tends to be present in greater amounts than the *trans* isomer. Isomerases can increase the reaction rate by lowering the isomerization energy.

Calculating isomerase kinetics from experimental data can be more difficult than for other enzymes because the use of product inhibition experiments is impractical. That is, isomerization is not an irreversible reaction since a reaction vessel will contain one substrate and one product so the typical simplified model for calculating reaction kinetics does not hold. There are also practical difficulties in determining the rate-determining step at high concentrations in a single isomerization. Instead, tracer perturbation can overcome these technical difficulties if there are two forms of the unbound enzyme. This technique uses isotope exchange to measure indirectly the interconversion of the free enzyme between its two forms. The radiolabeled substrate and product diffuse in a time-dependent manner. When the system reaches equilibrium the addition of unlabeled substrate perturbs or unbalances it. As equilibrium is established again, the radiolabeled substrate and product are tracked to determine energetic information.

The earliest use of this technique elucidated the kinetics and mechanism underlying the action of phosphoglucomutase, favoring the model of indirect transfer of phosphate with one intermediate and the direct transfer of glucose. This technique was then adopted to study the profile of proline racemase and its two states: the form which isomerizes L-proline and the other for D-proline. At high concentrations it was shown that the transition state in this interconversion is rate-limiting and that these enzyme forms may differ just in the protonation at the acidic and basic groups of the active site.

Nomenclature

Generally, "the names of isomerases are formed as "*substrate* isomerase" (for example, enoyl CoA isomerase), or as "*substrate type of isomerase*" (for example, phosphoglucomutase)."

Classification

Enzyme-catalyzed reactions each have a uniquely assigned classification number. Isomerase-catalyzed reactions have their own EC category: EC 5. Isomerases are further classified into six subclasses:

Racemases and Epimerases

This category (EC 5.1) includes (racemases) and epimerases). These isomerases invert stereochemistry at the target chiral carbon. Racemases act upon molecules with one chiral carbon for inversion of stereochemistry, whereas epimerases target molecules with multiple chiral carbons and act upon one of them. A molecule with only one chiral carbon has two enantiomeric forms, such as serine having the isoforms D-serine and L-serine differing only in the absolute configuration about the chiral carbon. A molecule with multiple chiral carbons has two forms at each chiral carbon. Isomerization at one chiral carbon of several yields epimers, which differ from one another in absolute configuration at just one chiral carbon. For example, D-glucose and D-mannose differ in configuration at just one chiral carbon. This class is further broken down by the group the enzyme acts upon:

Racemases and epimerases		
EC number	Description	Examples
EC 5.1.1	Acting on Amino Acids and Derivative	alanine racemase, methionine racemase
EC 5.1.2	Acting on Hydroxy Acids and Derivatives	lactate racemase, tartrate epimerase
EC 5.1.3	Acting on Carbohydrates and Derivatives	ribulose-phosphate 3-epimerase, UDP-glucose 4-epimerase
EC 5.1.99	Acting on Other Compounds	methylmalonyl CoA epimerase, hydantoin racemase

Cis-trans Isomerases

This category (EC 5.2) includes enzymes that catalyze the isomerization of cis-trans isomers. Alkenes and cycloalkanes may have cis-trans stereoisomers. These isomers are not distinguished by absolute configuration but rather by the position of substituent groups relative to a plane of reference, as across a double bond or relative to a ring structure. *Cis* isomers have substituent groups on the same side and *trans* isomers have groups on opposite sides.

This category is not broken down any further. All entries presently include:

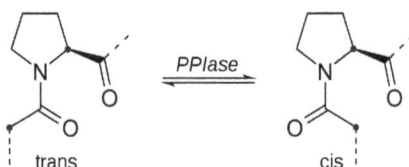

Conversion mediated by peptidylprolyl isomerase (PPIase).

Cis-trans isomerases	
EC number	Examples
EC 5.2.1.1	Maleate isomerase
EC 5.2.1.2	Maleylacetoacetate isomerase

EC 5.2.1.4	Maleylpyruvate isomerase
EC 5.2.1.5	Linoleate isomerase
EC 5.2.1.6	Furylfuramide isomerase
EC 5.2.1.8	Peptidylprolyl isomerase
EC 5.2.1.9	Farnesol 2-isomerase
EC 5.2.1.10	2-chloro-4-carboxymethylenebut-2-en-1,4-olide isomerase
EC 5.2.1.12	Zeta-carotene isomerase
EC 5.2.1.13	Prolycopene isomerase
EC 5.2.1.14	Beta-carotene isomerase

Intramolecular Oxidoreductases

This category (EC 5.3) includes intramolecular oxidoreductases. These isomerases catalyze the transfer of electrons from one part of the molecule to another. In other words, they catalyze the oxidation of one part of the molecule and the concurrent reduction of another part. Sub-categories of this class are:

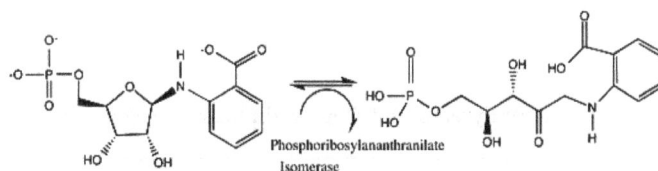

Reaction catalyzed by phosphoribosylanthranilate isomerase.

Intramolecular oxidoreductases		
EC number	Description	Examples
EC 5.3.1	Interconverting Aldoses and Ketoses	Triose-phosphate isomerase, Ribose-5-phosphate isomerase
EC 5.3.2	Interconverting Keto- and Enol-Groups	Phenylpyruvate tautomerase, Oxaloacetate tautomerase
EC 5.3.3	Transposing C=C Double Bonds	Steroid Delta-isomerase, L-dopachrome isomerase
EC 5.3.4	Transposing S-S Bonds	Protein disulfide-isomerase
EC 5.3.99	Other Intramolecular Oxidoreductases	Prostaglandin-D synthase, Allene-oxide cyclase

Intramolecular Transferases

This category (EC 5.4) includes intramolecular transferases (mutases). These isomerases catalyze the transfer of functional groups from one part of a molecule to another. Phosphotransferases (EC 5.4.2) were categorized as transferases (EC 2.7.5) with regeneration of donors until 1983. This sub-class can be broken down according to the functional group the enzyme transfers:

Reaction catalyzed by phosphoenolpyruvate mutase.

Intramolecular transferases		
EC number	Description	Examples
EC 5.4.1	Transferring Acyl Groups	Lysolecithin acylmutase, Precorrin-8X methylmutase
EC 5.4.2	Phosphotransferases (Phosphomutases)	Phosphoglucomutase, Phosphopentomutase
EC 5.4.3	Transferring Amino Groups	Beta-lysine 5,6-aminomutase, Tyrosine 2,3-aminomutase
EC 5.4.4	Transferring hydroxy groups	(hydroxyamino)benzene mutase, Isochorismate synthase
EC 5.4.99	Transferring Other Groups	Methylaspartate mutase, Chorismate mutase

Intramolecular Lyases

This category (EC 5.5) includes intramolecular lyases. These enzymes catalyze "reactions in which a group can be regarded as eliminated from one part of a molecule, leaving a double bond, while remaining covalently attached to the molecule." Some of these catalyzed reactions involve the breaking of a ring structure.

This category is not broken down any further. All entries presently include:

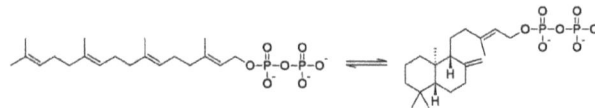

Reaction catalyzed by ent-Copalyl diphosphate synthase.

Intramolecular lyases	
EC number	Examples
EC 5.5.1.1	Muconate cycloisomerase
EC 5.5.1.2	3-carboxy-cis,cis-muconate cycloisomerase
EC 5.5.1.3	Tetrahydroxypteridine cycloisomerase
EC 5.5.1.4	Inositol-3-phosphate synthase
EC 5.5.1.5	Carboxy-cis,cis-muconate cyclase
EC 5.5.1.6	Chalcone isomerase
EC 5.5.1.7	Chloromuconate cycloisomerase
EC 5.5.1.8	(+)-bornyl diphosphate synthase
EC 5.5.1.9	Cycloeucalenol cycloisomerase
EC 5.5.1.10	Alpha-pinene-oxide decyclase
EC 5.5.1.11	Dichloromuconate cycloisomerase
EC 5.5.1.12	Copalyl diphosphate synthase
EC 5.5.1.13	Ent-copalyl diphosphate synthase
EC 5.5.1.14	Syn-copalyl-diphosphate synthase
EC 5.5.1.15	Terpentedienyl-diphosphate synthase
EC 5.5.1.16	Halimadienyl-diphosphate synthase
EC 5.5.1.17	(S)-beta-macrocarpene synthase
EC 5.5.1.18	Lycopene epsilon-cyclase
EC 5.5.1.19	Lycopene beta-cyclase
EC 5.5.1.20	Prosolanapyrone-III cycloisomerase
EC 5.5.1.n1	D-ribose pyranase

Mechanisms of Isomerases

Ring Expansion and Contraction via Tautomers

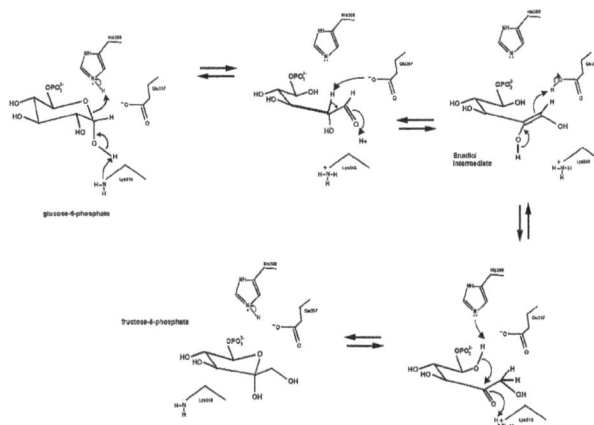

The isomerization of glucose-6-phosphate by glucose-6-phosphate isomerase.

A classic example of ring opening and contraction is the isomerization of glucose (an aldehyde with a six-membered ring) to fructose (a ketone with a five-membered ring). The conversion of D-glucose-6-phosphate to D-fructose-6-phosphate is catalyzed by glucose-6-phosphate isomerase, an intramolecular oxidoreductase. The overall reaction involves the opening of the ring to form an aldose via acid/base catalysis and the subsequent formation of a cis-endiol intermediate. A ketose is then formed and the ring is closed again.

Glucose-6-phosphate first binds to the active site of the isomerase. The isomerase opens the ring: its His388 residue protonates the oxygen on the glucose ring (and thereby breaking the O5-C1 bond) in conjunction with Lys518 deprotonating the C1 hydroxyl oxygen. The ring opens to form a straight-chain aldose with an acidic C2 proton. The C3-C4 bond rotates and Glu357 (assisted by His388) deprotonates C2 to form a double bond between C1 and C2. A cis-endiol intermediate is created and the C1 oxygen is protonated by the catalytic residue, accompanied by the deprotonation of the endiol C2 oxygen. The straight-chain ketose is formed. To close the fructose ring, the reverse of ring opening occurs and the ketose is protonated.

Epimerization

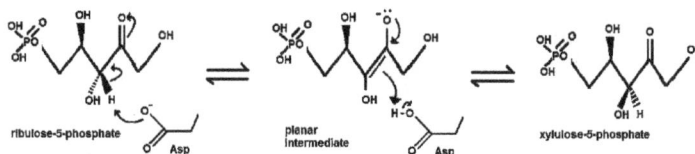

The conversion of ribulose-5-phosphate to xylulose-5-phosphate.

An example of epimerization is found in the Calvin cycle when D-ribulose-5-phosphate is converted into D-xylulose-5-phosphate by ribulose-phosphate 3-epimerase. The substrate and product differ only in stereochemistry at the third carbon in the chain. The underlying mechanism involves the deprotonation of that third carbon to form a reactive enolate intermediate. The enzyme's active site contains two Asp residues. After the substrate binds to the enzyme, the first Asp deprotonates the third carbon from one side of the molecule. This leaves a planar sp²-hybridized intermediate. The second Asp is located on the opposite side of the active side and it protonates the molecule, effectively adding a proton from the back side. These coupled steps invert stereochemistry at the third carbon.

Intramolecular Transfer

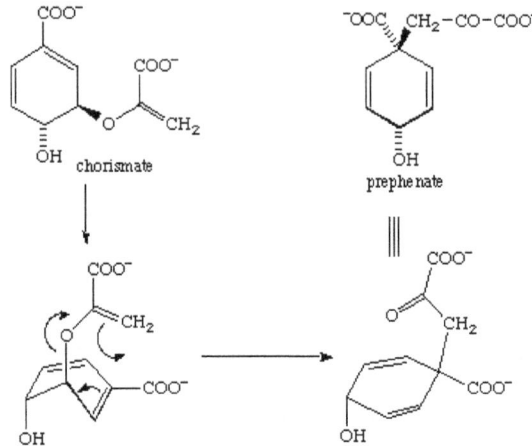

A proposed mechanism for chorismate mutase. Clark, T., Stewart, J.D. and Ganem,
B. Transition-state analogue inhibitors of chlorismate mutase.

Chorismate mutase is an intramolecular transferase and it catalyzes the conversion of chorismate to prephenate, used as a precursor for L-tyrosine and L-phenylalanine in some plants and bacteria. This reaction is a Claisen rearrangement that can proceed with or without the isomerase, though the rate increases 10⁶ fold in the presence of chorismate mutase. The reaction goes through a chair transition state with the substrate in a trans-diaxial position. Experimental evidence indicates that the isomerase selectively binds the chair transition state, though the exact mechanism of catalysis is not known. It is thought that this binding stabilizes the transition state through electrostatic effects, accounting for the dramatic increase in the reaction rate in the presence of the mutase or upon addition of a specifically-placed cation in the active site.

Intramolecular Oxidoreduction

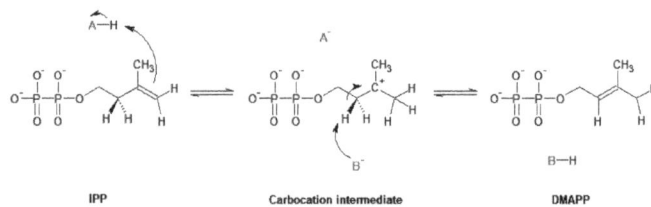

Conversion by IPP isomerase.

Isopentenyl-diphosphate delta isomerase type I (also known as IPP isomerase) is seen in cholesterol synthesis and in particular it catalyzes the conversion of isopentenyl diphosphate (IPP) to dimethylallyl diphosphate (DMAPP). In this isomerization reaction a stable carbon-carbon double bond is rearranged top create a highly electrophilic allylic isomer. IPP isomerase catalyzes this reaction by the stereoselective antarafacial transposition of a single proton. The double bond is protonated at

C4 to form a tertiary carbocation intermediate at C3. The adjacent carbon, C2, is deprotonated from the opposite face to yield a double bond. In effect, the double bond is shifted over.

Role of Isomerase in Human Disease

Isomerase plays a role in human disease. Deficiencies of this enzyme can cause disorders in humans.

Phosphohexose Isomerase Deficiency

Phosphohexose Isomerase Dificiency (PHI) is also known as phosphoglucose isomerase deficiency or Glucose-6-phosphate isomerase deficiency, and is a hereditary enzyme deficiency. PHI is the second most frequent erthoenzyopathy in glycolysis besides pyruvate kinase deficiency, and is associated with non-spherocytic haemolytic anaemia of variable severity. This disease is centered on the glucose-6-phosphate protein. This protein can be found in the secretion of some cancer cells. PHI is the result of a dimeric enzyme that catalyses the reversible interconversion of fructose-6-phosphate and gluose-6-phosphate.

PHI is a very rare disease with only 50 cases reported in literature to date.

Diagnosis is made on the basis of the clinical picture in association with biochemical studies revealing erythrocyte GPI deficiency (between 7 and 60% of normal) and identification of a mutation in the GPI gene by molecular analysis.

The deficiency of phosphohexose isomerase can lead to a condition referred to as hemolytic syndrome. As in humans, the hemolytic syndrome, which is characterized by a diminished erythrocyte number, lower hematocrit, lower hemoglobin, higher number of reticulocytes and plasma bilirubin concentration, as well as increased liver- and spleen-somatic indices, was exclusively manifested in homozygous mutants.

Triosephosphate Isomerase Deficiency

The disease referred to as triosephosphate isomerase deficiency (TPI), is a severe autosomal recessive inherited multisystem disorder of glycolyic metabolism. It is characterized by hemolytic anemia and neurodegeneration, and is caused by anaerobic metabolic dysfunction. This dysfunction results from a missense mutation that effects the encoded TPI protein. The most common mutation is the substitution of gene, Glu104Asp, which produces the most severe phenotype, and is responsible for approximately 80% of clinical TPI deficiency.

TPI deficiency is very rare with less than 50 cases reported in literature. Being an autosomal recessive inherited disease, TPI deficiency has a 25% recurrence risk in the case of heterozygous parents. It is a congenital disease that most often occurs with hemolytic anemia and manifests with jaundice. Most patients with TPI for Glu104Asp mutation or heterozygous for a TPI null allele and Glu104Asp have a life expectancy of infancy to early childhood. TPI patients with other mutations generally show longer life expectancy. To date, there are only two cases of individuals with TPI living beyond the age of 6. These cases involve two brothers from Hungary, one who did not develop neurological symptoms until the age of 12, and the older brother who has no neurological symptoms and suffers from anemia only.

Individuals with TPI show obvious symptoms after 6–24 months of age. These symptoms include: dystonia, tremor, dyskinesia, pyramidal tract signs, cardiomyopathy and spinal motor neuron involvement. Patients also show frequent respiratory system bacterial infections.

TPI is detected through deficiency of enzymatic activity and the build-up of dihyroxyacetone phosphate(DHAP), which is a toxic substrate, in erythrocytes. This can be detected through physical examination and a series of lab work. In detection, there is generally myopathic changes seen in muscles and chronic axonal neuropathy found in the nerves. Diagnosis of TPI can be confirmed through molecular genetics. Chorionic villus DNA analysis or analysis of fetal red cells can be used to detect TPI in antenatal diagnosis.

Treatment for TPI is not specific, but varies according to different cases. Because of the range of symptoms TPI causes, a team of specialist may be needed to provide treatment to a single individual. That team of specialists would consists of pediatricians, cardiologists, neurologists, and other healthcare professionals, that can develop a comprehensive plan of action.

Supportive measures such as red cell transfusions in cases of severe anaemia can be taken to treat TPI as well. In some cases, spleen removal (splenectomy) may improve the anaemia. There is no treatment to prevent progressive neurological impairment of any other non-haematological clinical manifestation of the diseases.

Industrial Applications

By far the most common use of isomerases in industrial applications is in sugar manufacturing. Glucose isomerase (also known as xylose isomerase) catalyzes the conversion of D-xylose and D-glucose to D-xylulose and D-fructose. Like most sugar isomerases, glucose isomerase catalyzes the interconversion of aldoses and ketoses.

The conversion of glucose to fructose is a key component of high-fructose corn syrup production. Isomerization is more specific than older chemical methods of fructose production, resulting in a higher yield of fructose and no side products. The fructose produced from this isomerization reaction is purer with no residual flavors from contaminants. High-fructose corn syrup is preferred by many confectionery and soda manufacturers because of the high sweetening power of fructose (twice that of sucrose), its relatively low cost and its inability to crystallize. Fructose is also used as a sweetener for use by diabetics. Major issues of the use of glucose isomerase involve its inactivation at higher temperatures and the requirement for a high pH (between 7.0 and 9.0) in the reaction environment. Moderately high temperatures, above 70 °C, increase the yield of fructose by at least half in the isomerization step. The enzyme requires a divalent cation such as Co^{2+} and Mg^{2+} for peak activity, an additional cost to manufacturers. Glucose isomerase also has a much higher affinity for xylose than for glucose, necessitating a carefully controlled environment.

The isomerization of xylose to xylulose has its own commercial applications as interest in biofuels has increased. This reaction is often seen naturally in bacteria that feed on decaying plant matter. Its most common industrial use is in the production of ethanol, achieved by the fermentation of xylulose. The use of hemicellulose as source material is very common. Hemicellulose contains xylan, which itself is composed of xylose in $\beta(1,4)$ linkages. The use of glucose isomerase very efficiently converts xylose to xylulose, which can then be acted upon by fermenting yeast. Overall, extensive research in genetic engineering has been invested into optimizing glucose isomerase and facilitating its recovery from industrial applications for reuse.

Glucose isomerase is able to catalyze the isomerization of a range of other sugars, including D-ribose, D-allose and L-arabinose. The most efficient substrates are those similar to glucose and xylose, having equatorial hydroxyl groups at the third and fourth carbons. The current model for the mechanism of glucose isomerase is that of a hydride shift based on X-ray crystallography and isotope exchange studies.

Membrane-associated Isomerases

Some isomerases associate with biological membranes as peripheral membrane proteins or anchored through a single transmembrane helix, for example isomerases with the thioredoxin domain, and certain prolyl isomerases.

Transferase

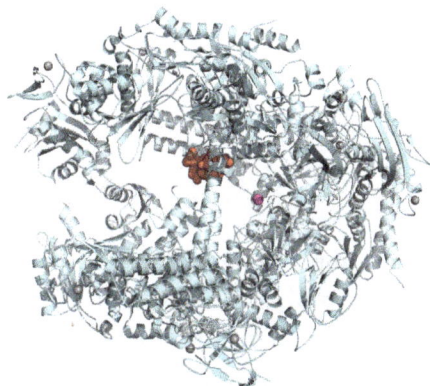

RNA polymerase from Saccharomyces cerevisiae complexed with α-amanitin (in red). Despite the use of the term "polymerase," RNA polymerases are classified as a form of nucleotidyl transferase.

A transferase is any one of a class of enzymes that enact the transfer of specific functional groups (e.g. a methyl or glycosyl group) from one molecule (called the donor) to another (called the acceptor). They are involved in hundreds of different biochemical pathways throughout biology, and are integral to some of life's most important processes.

Transferases are involved in myriad reactions in the cell. Three examples of these reactions are the activity of coenzyme A (CoA) transferase, which transfers thiol esters, the action of N-acetyltransferase, which is part of the pathway that metabolizes tryptophan, and the regulation of pyruvate dehydrogenase (PDH), which converts pyruvate to acetyl CoA. Transferases are also utilized during translation. In this case, an amino acid chain is the functional group transferred by a peptidyl transferase. The transfer involves the removal of the growing amino acid chain from the tRNA molecule in the A-site of the ribosome and its subsequent addition to the amino acid attached to the tRNA in the P-site.

Mechanistically, an enzyme that catalyzed the following reaction would be a transferase:

$$Xgroup + Y \xrightarrow{\text{transferase}} X + Ygroup$$

In the above reaction, X would be the donor, and Y would be the acceptor. "Group" would be the functional group transferred as a result of transferase activity. The donor is often a coenzyme.

Nomenclature

Systematic names of transferases are constructed in the form of "donor:acceptor grouptransferase." For example, methylamine:L-glutamate N-methyltransferase would be the standard naming convention for the transferase methylamine-glutamate N-methyltransferase, where methylamine is the donor, L-glutamate is the acceptor, and methyltransferase is the EC category grouping. This same action by the transferase can be illustrated as follows:

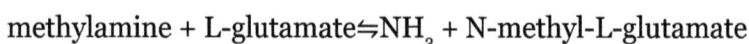

$$\text{methylamine} + \text{L-glutamate} \rightleftharpoons NH_3 + \text{N-methyl-L-glutamate}$$

However, other accepted names are more frequently used for transferases, and are often formed as "acceptor grouptransferase" or "donor grouptransferase." For example, a DNA methyltransferase is a transferase that catalyzes the transfer of a methyl group to a DNA acceptor. In practice, many molecules are not referred to using this terminology due to more prevalent common names. For example, RNA Polymerase is the modern common name for what was formerly known as RNA nucleotidyltransferase, a kind of nucleotidyl transferase that transfers nucleotides to the 3' end of a growing RNA strand. In the EC system of classification, the accepted name for RNA Polymerase is DNA-directed RNA polymerase.

Classification

Described primarily based on the type of biochemical group transferred, transferases can be divided into ten categories (based on the EC Number classification). These categories comprise over 450 different unique enzymes. In the EC numbering system, transferases have been given a classification of EC2. Hydrogen is not considered a functional group when it comes to transferase targets; instead, hydrogen transfer is included under oxidoreductases, due to electron transfer considerations.

Classification of transferases into subclasses		
EC number	Examples	Group(s) transferred
EC 2.1	methyltransferase and formyltransferase	single-carbon groups
EC 2.2	transketolase and transaldolase	aldehyde or ketone groups
EC 2.3	acyltransferase	acyl groups or groups that become alkyl groups during transfer
EC 2.4	glycosyltransferase, hexosyltransferase, and pentosyltransferase	glycosyl groups, as well as hexoses and pentoses
EC 2.5	riboflavin synthase and chlorophyll synthase	alkyl or aryl groups, other than methyl groups
EC 2.6	transaminase, and oximinotransferase	nitrogenous groups
EC 2.7	phosphotransferase, polymerase, and kinase	phosphorus-containing groups; subclasses are based on the acceptor (e.g. alcohol, carboxyl, etc.)
EC 2.8	sulfurtransferase and sulfotransferase	sulfur-containing groups
EC 2.9	selenotransferase	selenium-containing groups
EC 2.10	molybdenumtransferase and tungstentransferase	molybdenum or tungsten

Reactions

EC 2.1: Single Carbon Transferases

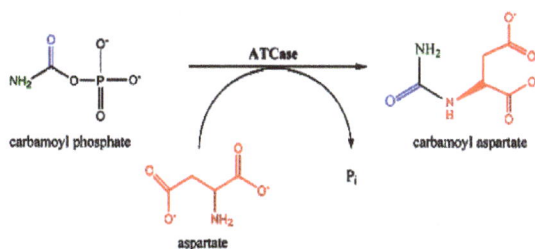

Reaction involving aspartate transcarbamylase

EC 2.1 includes enzymes that transfer single-carbon groups. This category consists of transfers of methyl, hydroxymethyl, formyl, carboxy, carbamoyl, and amido groups. Carbamoyltransferases, as an example, transfer a carbamoyl group from one molecule to another. Carbamoyl groups follow the formula NH_2CO. In ATCase such a transfer is written as,

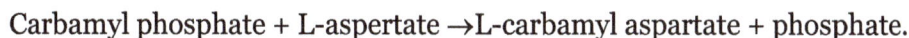

Carbamyl phosphate + L-aspertate →L-carbamyl aspartate + phosphate.

EC 2.2: Aldehyde and Ketone Transferases

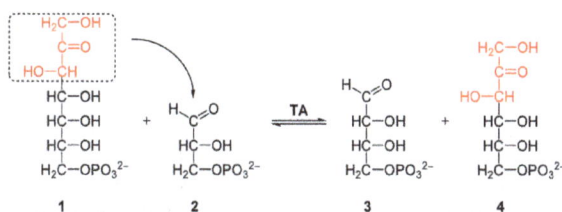

The reaction catalyzed by transaldolase

Enzymes that transfer aldehyde or ketone groups and included in EC 2.2. This category consists of various transketolases and transaldolases. Transaldolase, the namesake of aldehyde transferases, is an important part of the pentose phosphate pathway. The reaction it catalyzes consists of a transfer of a dihydroxyacetone functional group to Glyceraldehyde 3-phosphate (also known as G3P). The reaction is as follows:

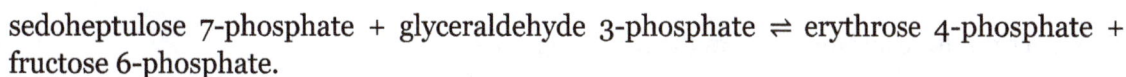

sedoheptulose 7-phosphate + glyceraldehyde 3-phosphate ⇌ erythrose 4-phosphate + fructose 6-phosphate.

EC 2.3: Acyl Transferases

Transfer of acyl groups or acyl groups that become alkyl groups during the process of being transferred are key aspects of EC 2.3. Further, this category also differentiates between amino-acyl and non-amino-acyl groups. Peptidyl transferase is a ribozyme that facilitates formation of peptide bonds during translation. As an aminoacyltransferase, it catalyzes the transfer of a peptide to an aminoacyl-tRNA, following this reaction:

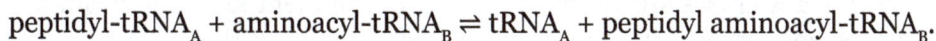

peptidyl-$tRNA_A$ + aminoacyl-$tRNA_B$ ⇌ $tRNA_A$ + peptidyl aminoacyl-$tRNA_B$.

EC 2.4: Glycosyl, Hexosyl and Pentosyl Transferases

EC 2.4 includes enzymes that transfer glycosyl groups, as well as those that transfer hexose and pentose. Glycosyltransferase is a subcategory of EC 2.4 transferases that is involved in biosynthesis of disaccharides and polysaccharides through transfer of monosaccharides to other molecules. An example of a prominent glycosyltransferase is lactose synthase which is a dimer possessing two protein subunits. Its primary action is to produce lactose from glucose and UDP-galactose. This occurs via the following pathway:

$$\text{UDP-}\beta\text{-D-galactose} + \text{D-glucose} \rightleftharpoons \text{UDP} + \text{lactose.}$$

EC 2.5: Alkyl and Aryl Transferases

EC 2.5 relates to enzymes that transfer alkyl or aryl groups, but does not include methyl groups. This is in contrast to functional groups that become alkyl groups when transferred, as those are included in EC 2.3. EC 2.5 currently only possesses one sub-class: Alkyl and aryl transferases. Cysteine synthase, for example, catalyzes the formation of acetic acids and cysteine from O_3-acetyl-L-serine and hydrogen sulfide:

$$O_3\text{-acetyl-L-serine} + H_2S \rightleftharpoons \text{L-cysteine} + \text{acetate.}$$

EC 2.6: Nitrogenous Transferases

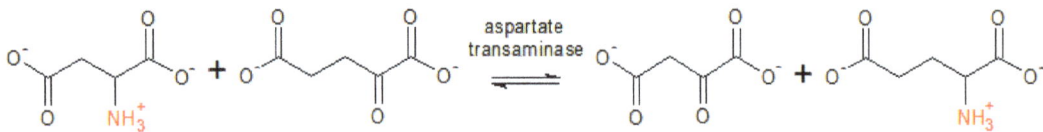

Aspartate aminotransferase can act on several different amino acids

The grouping consistent with transfer of nitrogenous groups is EC 2.6. This includes enzymes like transaminase (also known as "aminotransferase"), and a very small number of oximinotransferases and other nitrogen group transferring enzymes. EC 2.6 previously included amidinotransferase but it has since been reclassified as a subcategory of EC 2.1 (single-carbon transferring enzymes). In the case of aspartate transaminase, which can act on tyrosine, phenylalanine, and tryptophan, it reversibly transfers an amino group from one molecule to the other.

The reaction, for example, follows the following order:

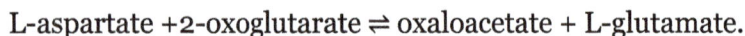

$$\text{L-aspartate} + 2\text{-oxoglutarate} \rightleftharpoons \text{oxaloacetate} + \text{L-glutamate.}$$

EC 2.7: Phosphorus Transferases

While EC 2.7 includes enzymes that transfer phosphorus-containing groups, it also includes nuclotidyl transferases as well. Sub-category phosphotransferase is divided up in categories based on the type of group that accepts the transfer. Groups that are classified as phosphate acceptors include: alcohols, carboxy groups, nitrogenous groups, and phosphate groups. Further constituents of this subclass of transferases are various kinases. A prominent kinase is cyclin-dependent kinase (or CDK), which comprises a sub-family of protein kinases. As their name implies, CDKs are heavily dependent on specific cyclin molecules for activation. Once combined, the CDK-cyclin complex is capable of enacting its function within the cell cycle.

The reaction catalyzed by CDK is as follows:

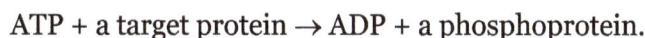

> ATP + a target protein → ADP + a phosphoprotein.

EC 2.8: Sulfur Transferases

Transfer of sulfur-containing groups is covered by EC 2.8 and is subdivided into the subcategories of sulfurtransferases, sulfotransferases, and CoA-transferases, as well as enzymes that transfer alkylthio groups. A specific group of sulfotransferases are those that use PAPS as a sulfate group donor. Within this group is alcohol sulfotransferase which has a broad targeting capacity. Due to this, alcohol sulfotransferase is also known by several other names including "hydroxysteroid sulfotransferase," "steroid sulfokinase," and "estrogen sulfotransferase." Decreases in its activity has been linked to human liver disease. This transferase acts via the following reaction:

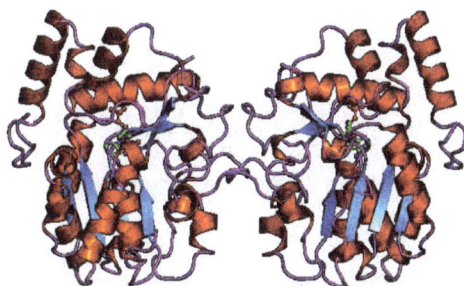

> 3'-phosphoadenylyl sulfate + an alcohol ⇌ adenosine 3',5'bisphosphate + an alkyl sulfate.

Ribbon diagram of a variant structure of estrogen sulfotransferase (PDB 1aqy EBI)

EC 2.9: Selenium Transferases

EC 2.9 includes enzymes that transfer selenium-containing groups. This category only contains two transferases, and thus is one of the smallest categories of transferase. Selenocysteine synthase, which was first added to the classification system in 1999, converts seryl-tRNA(Sec UCA) into selenocysteyl-tRNA(Sec UCA).

EC 2.10: Metal Transferases

The category of EC 2.10 includes enzymes that transfer molybdenum or tungsten-containing groups. However, as of 2011, only one enzyme has been added: molybdopterin molybdotransferase. This enzyme is a component of MoCo biosynthesis in *Escherichia coli*. The reaction it catalyzes is as follows:

> adenylyl-molybdopterin + molybdate → molybdenum cofactor + AMP.

Role in Histo-blood Group

The A and B transferases are the foundation of the human ABO blood group system. Both A and B transferases are glycosyltransferases, meaning they transfer a sugar molecule onto an H-antigen.

This allows H-antigen to synthesize the glycoprotein and glycolipid conjugates that are known as the A/B antigens. The full name of A transferase is alpha 1-3-N-acetylgalactosaminyltransferase and its function in the cell is to add N-acetylgalactosamine to H-antigen, creating A-antigen. The full name of B transferase is alpha 1-3-galactosyltransferase, and its function in the cell is to add a galactose molecule to H-antigen, creating B-antigen.

It is possible for *Homo sapiens* to have any of four different blood types: Type A (express A antigens), Type B (express B antigens), Type AB (express both A and B antigens) and Type O (express neither A nor B antigens). The gene for A and B transferases is located on chromosome 9. The gene contains seven exons and six introns and the gene itself is over 18kb long. The alleles for A and B transferases are extremely similar. The resulting enzymes only differ in 4 amino acid residues. The differing residues are located at positions 176, 235, 266, and 268 in the enzymes.

Deficiencies

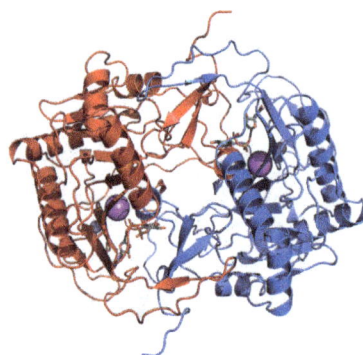

A deficiency of this transferase, E. coli galactose-1-phosphate
uridyltransferase is a known cause of galactosemia

Transferase deficiencies are at the root of many common illnesses. The most common result of a transferase deficiency is a buildup of a cellular product.

SCOT Deficiency

Succinyl-CoA:3-ketoacid CoA transferase deficiency (or SCOT deficiency) leads to a buildup of ketones. Ketones are created upon the breakdown of fats in the body and are an important energy source. Inability to utilize ketones leads to intermittent ketoacidosis, which usually first manifests during infancy. Disease sufferers experience nausea, vomiting, inability to feed, and breathing difficulties. In extreme cases, ketoacidosis can lead to coma and death. The deficiency is caused by mutation in the gene OXCT1. Treatments mostly rely on controlling the diet of the patient.

CPT-II Deficiency

Carnitine palmitoyltransferase II deficiency (also known as CPT-II deficiency) leads to an excess long chain fatty acids, as the body lacks the ability to transport fatty acids into the mitochondria to be processed as a fuel source. The disease is caused by a defect in the gene CPT2. This deficiency will present in patients in one of three ways: lethal neonatal, severe infantile hepatocardiomuscular, and myopathic form. The myopathic is the least severe form of the deficiency and can manifest at any point in the lifespan of the patient. The other two forms appear in infancy. Common symptoms of the lethal neonatal form and the severe infantile forms are liver failure, heart problems, seizures and death. The myopathic form is characterized by muscle pain and weakness following vigorous exercise. Treatment generally includes dietary modifications and carnitine supplements.

Galactosemia

Galactosemia results from an inability to process galactose, a simple sugar. This deficiency occurs when the gene for galactose-1-phosphate uridylyltransferase (GALT) has any number of mutations, leading to a deficiency in the amount of GALT produced. There are two forms of Galactosemia: classic and Duarte. Duarte galactosemia is generally less severe than classic galactosemia and is caused by a deficiency of galactokinase. Galactosemia renders infants unable to process the sugars in breast milk, which leads to vomiting and anorexia within days of birth. Most symptoms of the disease are caused by a buildup of galactose-1-phosphate in the body. Common symptoms include liver failure, sepsis, failure to grow, and mental impairment, among others. Buildup of a second toxic substance, galactitol, occurs in the lenses of the eyes, causing cataracts. Currently, the only available treatment is early diagnosis followed by adherence to a diet devoid of lactose, and prescription of antibiotics for infections that may develop.

Choline Acetyltransferase Deficiencies

Choline acetyltransferase (also known as ChAT or CAT) is an important enzyme which produces the neurotransmitter acetylcholine. Acetylcholine is involved in many neuropsychic functions such as memory, attention, sleep and arousal. The enzyme is globular in shape and consists of a single amino acid chain. ChAT functions to transfer an acetyl group from acetyl co-enzyme A to choline in the synapses of nerve cells and exists in two forms: soluble and membrane bound. The ChAT gene is located on chromosome 10.

Alzheimer's Disease

Decreased expression of ChAT is one of the hallmarks of Alzheimer's disease. Patients with Alzheimer's disease show a 30 to 90% reduction in activity in several regions of the brain, including the temporal lobe, the parietal lobe and the frontal lobe. However, ChAT deficiency is not believed to be the main cause of this disease.

Amyotrophic Lateral Sclerosis (ALS or Lou Gehrig's Disease)

Patients with ALS show a marked decrease in ChAT activity in motor neurons in the spinal cord and brain. Low levels of ChAT activity are an early indication of the disease and are detectable long before motor neurons begin to die. This can even be detected before the patient is symptomatic.

Huntington's Disease

Patients with Huntington's also show a marked decrease in ChAT production. Though the specific cause of the reduced production is not clear, it is believed that the death of medium-sized motor neurons with spiny dendrites leads to the lower levels of ChAT production.

Schizophrenia

Patients with Schizophrenia also exhibit decreased levels of ChAT, localized to the mesopontine tegment of the brain and the nucleus accumbens, which is believed to correlate with the decreased cognitive functioning experienced by these patients.

Sudden Infant Death Syndrome (SIDS)

Recent studies have shown that SIDS infants show decreased levels of ChAT in both the hypothalamus and the striatum. SIDS infants also display fewer neurons capable of producing ChAT in the vagus system. These defects in the medulla could lead to an inability to control essential autonomic functions such as the cardiovascular and respiratory systems.

Congenital Myasthenic Syndrome (CMS)

CMS is a family of diseases that are characterized by defects in neuromuscular transmission which leads to recurrent bouts of apnea (inability to breathe) that can be fatal. ChAT deficiency is implicated in myasthenia syndromes where the transition problem occurs presynaptically. These syndromes are characterized by the patients' inability to resynthesize acetylcholine.

Uses in Biotechnology

Terminal Transferases

Terminal transferases are transferases that can be used to label DNA or to produce plasmid vectors. It accomplishes both of these tasks by adding deoxynucleotides in the form of a template to the downstream end or 3' end of an existing DNA molecule. Terminal transferase is one of the few DNA polymerases that can function without an RNA primer.

Glutathione Transferases

The family of glutathione transferases (GST) is extremely diverse, and therefore can be used for a number of biotechnological purposes. Plants use glutathione transferases as a means to segregate toxic metals from the rest of the cell. These glutathione transferases can be used to create biosensors to detect contaminants such as herbicides and insecticides. Glutathione transferases are also used in transgenic plants to increase resistance to both biotic and abiotic stress. Glutathione transferases are currently being explored as targets for anti-cancer medications due to their role in drug resistance. Further, glutathione transferase genes have been investigated due to their ability to prevent oxidative damage and have shown improved resistance in transgenic cultigens.

Rubber Transferases

Currently the only available commercial source of natural rubber is the Hevea plant (Hevea brasiliensis). Natural rubber is superior to synthetic rubber in a number of commercial uses. Efforts are being made to produce transgenic plants capable of synthesizing natural rubber, including tobacco and sunflower. These efforts are focused on sequencing the subunits of the rubber transferase enzyme complex in order to transfect these genes into other plants.

Membrane-associated Transferases

Many transferases associate with biological membranes as peripheral membrane proteins or anchored to membranes through a single transmembrane helix, for example numerous glycosyltransferases in Golgi apparatus. Some others are multi-span transmembrane proteins, for example certain Oligosaccharyltransferases or microsomal glutathione S-transferase from MAPEG family.

Lyases

Lyase is an enzyme catalytically aiding in breaking various chemical bonds by means of an "elimination" reaction, other than hydrolysis and oxidation. This reaction often results in the formation of a new cyclic structure or a new double bond, and a reverse reaction called a "Michael addition" might also possibly happen under the catalysis of lyase. To obtain either a double bond or a new ring, lyase acts upon the single substrate and a molecule is eliminated. Lyases are different from other enzymes for only one substrate is required for the reaction in one direction, but two substrates are essential for the reverse reaction. Lyases can be commonly observed in the reactions of the Citric Acid Cycle (Krebs cycle) and in glycolysis. In glycolysis, aldolase could readily and reversibly degrade fructose 1,6-bisphosphate into the products glyceraldehyde 3-phosphate and dihydroxyacetone phosphate, which is an example of a lyase cleaving carbon-carbon bonds. Lyase works without the necessary requirements for cofactor recycling and gives an absolute stereospecificity with a theoretical yield of 100%, being much more efficient compared with enantiomeric resolutions of only 50% productive rate. Therefore, considerable researches have been addicted to the exploration of lyases as biocatalysts to synthesize optically active compounds, which have also been already found application in a few large commercial processes. Lyases are systematically named as "substrate group-lyase", such as decarboxylase, dehydratase, aldolase, etc.

Classification

EC 4: Lyases

1. EC 4.1: Cleave carbon-carbon bonds

2. EC 4.2: Cleave carbon-oxygen bonds

3. EC 4.3: Cleave carbon-nitrogen bonds

4. EC 4.4: Cleave carbon-sulfur bonds

5. EC 4.5: Cleave carbon-halide bonds

6. EC 4.6: Cleave phosphorus-oxygen bonds

7. EC 4.99: A group of other lyases

In the EC number classification of enzymes, EC 4 could represent lyases, which can be further classified into seven subclasses. Lyases in EC 4.1 cleave carbon-carbon bonds, and include decarboxylases (EC 4.1.1), aldehyde lyases (EC 4.1.2) facilitating the reverse reaction of aldol condensations, oxo acid lyases (EC 4.1.3) that catalyzes the cleavage of many 3-hydroxy acids, and others (EC 4.1.99). EC 4.2 contains a group of lyases that break carbon-oxygen bonds, such as dehydratases. Hydro-lyases being a part of carbon-oxygen lyases could facilitate the cleavage of C-O bonds by the elimination of water. Some other carbon-oxygen lyases promote the elimination of a phosphate or the removal of an alcohol from a polysaccharide. Lyases cleaving carbon-nitrogen bonds are sorted into EC 4.3. They could release ammonia with powerful cleaving ability and simultaneously form a double bond or ring. Some of these enzymes can also help to eliminate an amine or amide group. EC 4.4 represents lyases that split carbon-sulfur bonds, which could eliminate or substitute dihydrogen sulfide (H_2S) from a reaction. Carbon-halide bonds cleaving enzymes are lyases in EC 4.5 and that utilize an action mode that removes hydrochloric acid from a synthetic pesticide dichloro-diphenyl-trichloroethane (DDT). EC 4.6 comprises lyases fracturing phosphorus-oxygen bonds, like adenylyl cyclase and guanylyl cyclase, and they eliminate diphosphate from nucleotide triphosphates. EC 4.99 is a group of other lyases.

Substrate Specificity

Narrow substrate specificity is usually considered to be a drawback for the commercialization of an enzyme in that it greatly restricts the flexibility of an enzyme as an assistant in the production of related compounds. Lyases are generally, but not always, found with narrow substrate specificity. Most hydratases and ammonia-lyases indeed possess quite narrow substrate specificity, while the substrate specificity for aldolases, decarboxylases and oxynitrilases is much broader. It is noteworthy here that the substrate specificity of a specific lyase varies depending on its source. However, it is not an absolute prerequisite for enzymes to own unrestricted substrate specificity for their commercial exploitation. In fact, there are several of the lyases in commercial use bearing a rather narrow substrate spectrum.

Cofactor Requirements

The commercial potential of enzymes can be severely limited by the requirement for expensive cofactors. Since the addition catalyzed by lyase does not implicate a mere oxidation or reduction, it is not an essential requirement for cofactors. However, up to now, most of the lyases identified do require cofactors, which are involved in stabilization of reaction intermediates, polarization of the substrate, substrate binding, temporary binding of the nucleophile, and so on. The majority of these cofactors are not very expensive, and covalently bound to the enzyme. Thereby, the cofactors of lyases do not establish a barrier to their commercialization. The requirements for cofactors of lyases are varied according to their different sources.

Ligase

Ligase is any one of a class of about 50 enzymes that catalyze reactions involving the conservation of chemical energy and provide a couple between energy-demanding synthetic processes and energy-yielding breakdown reactions. They catalyze the joining of two molecules, deriving the needed energy from the cleavage of an energy-rich phosphate bond (in many cases, by the simultaneous conversion of adenosine triphosphate [ATP] to adenosine diphosphate [ADP]). A ligase catalyzing the formation of a carbon-oxygen bond between an amino acid and transfer RNA is called amino acid–RNA ligase. Carbon–nitrogen (C–N) bonds are formed by the action of such enzymes as amide synthetases and peptide synthetases.

Working Principle of Enzymes

Enzymes help reduce the activation energy of the complex molecules in the reaction. The following steps simplify how an enzyme works to speed up a reaction:

Step 1: Each enzyme has an 'active site' which is where one of the substrate molecules can bind to. Thus, an enzyme- substrate complex is formed.

Step 2: This enzyme-substrate molecule now reacts with the second substrate to form the product and the enzyme is liberated as the second product.

There are two theories to describe enzyms:

Theory 1: Lock and Key Hypothesis

This is the most accepted of the theories of enzyme action.

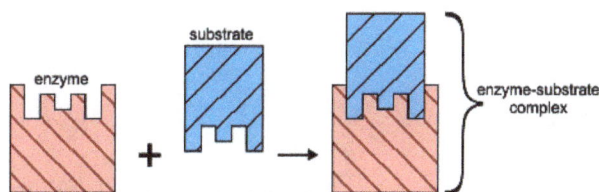

This theory states that the substrate fits exactly into the active site of the enzyme to form an enzyme-substrate complex. This model also describes why enzymes are so specific in their action because they are specific to the substrate molecules.

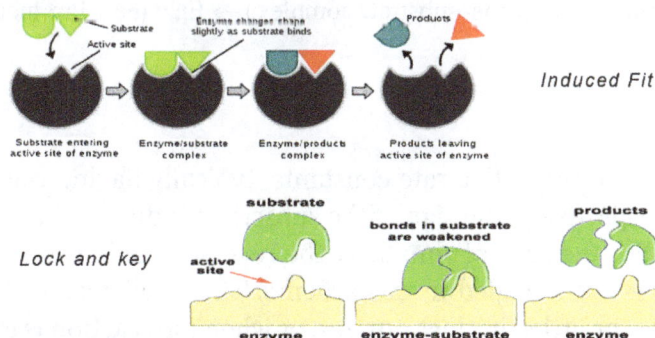

Theory 2: Induced Fit Hypothesis

This is similar to the lock and key hypothesis. It says that the shape of the enzyme molecule changes as it gets closer to the substrate molecule in such a way that the substrate molecule fits exactly into the active site of the enzyme.

Factors Affecting Enzyme Activity in the Cell

- Concentration of Enzymes and Substrates: The rate of reaction increases with increasing substrate concentration up to a point, beyond which any further increase in substrate concentration produces no significant change in reaction rate. This occurs because after a certain concentration of the substrate, all the active sites on the enzyme are full and no further reaction can occur.

- Temperature: With the increase in temperature, the enzyme activity increases because of the increase in kinetic energy of the molecules. There is an optimum level when the enzymes work at the best and maximum. This temperature is often the normal body temperature of the body. When the temperature increases beyond a certain limit, enzymes, which are actually made up of proteins, begin to disintegrate and the rate of reaction slows down.

- pH: Enzymes are very sensitive to changes in the pH and work in a very small window of permissible pH levels. Below or above the optimum pH level, there is a risk of the enzymes disintegrating and thereby the reaction slows down.

- Inhibitors: Presence of certain substances that inhibit the action of a particular enzyme. This occurs when the inhibiting substance attaches itself to the active site of the enzyme thereby preventing the substrate attachment and slows down the process.

Enzyme Kinetics

It is established that enzymes form a bound complex to their reactants (i.e., substrates) during the course of their catalysis and prior to the release of products. This can be simply illustrated, using the mechanism based on that of Michaelis and Menten for a one-substrate reaction, by the reaction sequence:

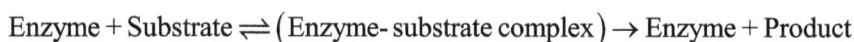

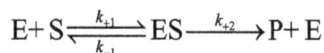

$$\text{Enzyme} + \text{Substrate} \rightleftharpoons \left(\text{Enzyme-substrate complex}\right) \rightarrow \text{Enzyme} + \text{Product}$$

$$E + S \underset{k_{-1}}{\overset{k_{+1}}{\rightleftharpoons}} ES \overset{k_{+2}}{\longrightarrow} P + E$$

where k_{+1}, k_{-1} and k_{+2} are the respective rate constants, typically having values of 10^5 - 10^8 M^{-1} s^{-1}, 1 - 10^4 s^{-1} and 1 - 10^5 s^{-1} respectively; the sign of the subscripts indicating the direction in which the rate constant is acting. For the sake of simplicity the reverse reaction concerning the conversion of product to substrate is not included in this scheme. This is allowable at the beginning of the reaction when there is no, or little, product present, or when the reaction is effectively irreversible. The rate of reaction (v) is the rate at which the product is formed.

$$v = \frac{d[P]}{dt} = k_{+2}[ES]$$

where [ ] indicates the molar concentration of the material enclosed (i.e., [ES] is the concentration of the enzyme-substrate complex). The rate of change of the concentration of the enzyme-substrate complex equals the rate of its formation minus the rate of its breakdown, forwards to give product or backwards to regenerate substrate.

therefore:

$$\frac{d[\text{ES}]}{dt} = k_{+1}[\text{E}][S] - (k_{-1} + k_{+2})[\text{ES}]$$

During the course of the reaction, the total enzyme at the beginning of the reaction ($[\text{E}]_0$, at zero time) is present either as the free enzyme ($[\text{E}]$) or the ES complex ($[\text{ES}]$).

i.e. $[\text{E}]_0 = [\text{E}] + [\text{ES}]$

therefore:

$$\frac{d[\text{ES}]}{dt} = k_{+1}([\text{E}]_0 - [\text{ES}])[S] - (k_{-1} + k_{+2})[\text{ES}]$$

Gathering terms together,

$$\frac{d[\text{ES}]}{dt} = k_{+1}[\text{E}]_0[S] - k_{+1}[\text{ES}][S] - -(k_{-1} + k_{+2})[\text{ES}]$$

$$\frac{d[\text{ES}]}{dt} = k_{+1}[\text{E}]_0[S] - (k_{+1}[S] + k_{-1} + k_{+2})[\text{ES}]$$

This gives:

$$\frac{\dfrac{d[\text{ES}]}{dt}}{k_{+1}[S] + k_{-1} + k_{+2}} + [\text{ES}] = \frac{k_{+1}[\text{E}]_0[S]}{k_{+1}[S] + k_{-1} + k_{+2}}$$

The differential equation 1.5 is difficult to handle, but may be greatly simplified if it can be assumed that the left hand side is equal to [ES] alone. This assumption is valid under the sufficient but unnecessarily restrictive steady state approximation that the rate of formation of ES equals its rate of disappearance by product formation and reversion to substrate (i.e., d[ES]/dt is zero). It is additionally valid when the condition is valid:

$$\frac{\dfrac{d[\text{ES}]}{dt}}{k_{+1}[S] + k_{-1} + k_{+2}} \ll [\text{ES}]$$

This occurs during a substantial part of the reaction time-course over a wide range of kinetic rate constants and substrate concentrations and at low to moderate enzyme concentrations. The variation in [ES], d[ES]/dt, [S] and [P] with the time-course of the reaction is shown in Figure, where it may be seen that the simplified equation is valid throughout most of the reaction.

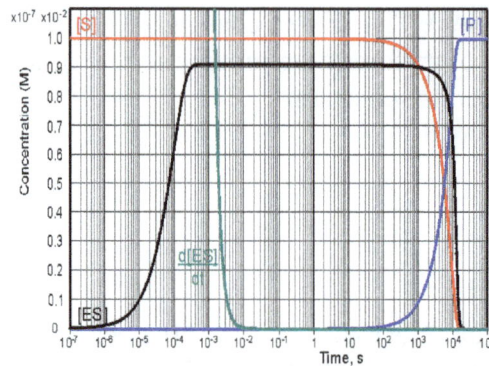

Figure: Computer simulation of the progress curves of d[ES]/dt (0 - 10^{-7} M scale), [ES] (0 - 10^{-7} M scale), [S] (0 - 10^{-2} M scale) and [P] (0 - 10^{-2} M scale) for a reaction obeying simple Michaelis-Menten kinetics with $k_{+1} = 10^6$ $M^{-1}s^{-1}$, $k_{-1} = 1000$ s^{-1}, $k_{+2} = 10$ s^{-1}, $[E]_0 = 10^{-7}$ M and $[S]_0 = 0.01$ M.

The simulation shows three distinct phases to the reaction time-course, an initial transient phase which lasts for about a millisecond followed by a longer steady state phase of about 30 minutes when [ES] stays constant but only a small proportion of the substrate reacts. This is followed by the final phase, taking about 6 hours during which the substrate is completely converted to product.

$$\dfrac{\dfrac{d[\text{ES}]}{dt}}{k_{+1}[S] + k_{-1} + k_{+2}}$$ is much less than [ES] during both of the latter two phases.

The Michaelis-Menten equation (below) is simply derived from equations $v = \dfrac{d[P]}{dt} = k_{+2}[\text{ES}]$ and

$$\dfrac{\dfrac{d[\text{ES}]}{dt}}{k_{+1}[S] + k_{-1} + k_{+2}} + [\text{ES}] = \dfrac{k_{+1}[\text{E}]_0[S]}{k_{+1}[S] + k_{-1} + k_{+2}},$$ by substituting K_m for $\dfrac{k_{-1} + k_{+2}}{k_{+1}}$. K_m is known as the Mi-

chaelis constant with a value typically in the range 10^{-1} - 10^{-5} M. When $k_{+2} << k_{-1}$, K_m equals the dissociation constant (k_{-1}/k_{+1}) of the enzyme substrate complex.

$$v = k_{+2}[\text{ES}] = \dfrac{k_{+2}[\text{E}]_0[S]}{[S] + K_m}$$

Or, more simply

$$v = \dfrac{V_{\max}[S]}{[S] + K_m}$$

Where V_{max} is the maximum rate of reaction, which occurs when the enzyme is completely saturated with substrate (i.e., when [S] is very much greater than K_m, V_{max} equals $k_{+2}[E]_0$, as the maximum value [ES] can have is $[E]_0$ when $[E]_0$ is less than $[S]_0$). Equation 1.8 may be rearranged to show the dependence of the rate of reaction on the ratio of [S] to K_m,

$$v = \dfrac{V_{\max}}{1 + \dfrac{K_m}{[S]}}$$

and the rectangular hyperbolic nature of the relationship, having asymptotes at v = V_{max} and [S] = $-K_m$,

$$(V_{max}-v)(K_m+[S])=V_{max}K_m$$

The substrate concentration in these equations is the actual concentration at the time and, in a closed system, will only be approximately equal to the initial substrate concentration ([S]$_o$) during the early phase of the reaction. Hence, it is usual to use these equations to relate the initial rate of reaction to the initial, and easily predetermined, substrate concentration. This also avoids any problem that may occur through product inhibition or reaction reversibility.

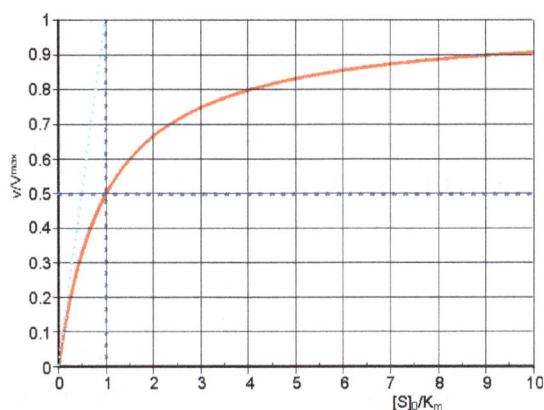

Figure: A normalised plot of the initial rate (v$_o$) against initial substrate concentration ([S]$_o$) for a reaction obeying the Michaelis-Menten kinetics.

The plot has been normalised in order to make it more generally applicable by plotting the relative initial rate of reaction (v$_o$/V_{max}) against the initial substrate concentration relative to the Michaelis constant ([S]$_o$/K_m, more commonly referred to as b, the dimensionless substrate concentration). The curve is a rectangular hyperbola with asymptotes at v$_o$ = V_{max} and [S]$_o$ = $-K_m$. The tangent to the curve at the origin goes through the point (v$_o$ = V_{max}), ([S]$_o$ = K_m). The ratio V_{max}/K_m is an important kinetic parameter which describes the relative specificity of a fixed amount of the enzyme for its substrate (more precisely defined in terms of k_{cat}/K_m). The substrate concentration, which gives a rate of half the maximum reaction velocity, is equal to the K_m.

It has been established that few enzymes follow the Michaelis-Menten equation over a wide range of experimental conditions. However, it remains by far the most generally applicable equation for describing enzymic reactions. Indeed it can be realistically applied to a number of reactions which have a far more complex mechanism than the one described here. In these cases K_m remains an important quantity, characteristic of the enzyme and substrate, corresponding to the substrate concentration needed for half the enzyme molecules to bind to the substrate (and, therefore, causing the reaction to proceed at half its maximum rate) but the precise kinetic meaning derived earlier may not hold and may be misleading. In these cases the K$_m$ is likely to equal a much more complex relationship between the many rate constants involved in the reaction scheme. It remains independent of the enzyme and substrate concentrations and indicates the extent of binding between the enzyme and its substrate for a given substrate concentration, a lower K$_m$ indicating a greater extent of binding. V_{max} clearly depends on the enzyme concentration and for some, but not all, enzymes may be largely independent of the specific substrate used. K_m and V_{max} may both be influenced by the charge and conformation of the protein and substrate(s) which are determined by pH, temperature, ionic strength and other factors. It is often preferable to substitute k_{cat} for k_{+2}, where $V_{max} = k_{cat}[E]_0$, as the precise meaning of k_{+2}, above, may also be misleading. k_{cat} is also known as the turnover number as it represents the maximum number of substrate molecules that the enzyme can 'turn over' to product in a set time (e.g. the turnover numbers of a-amylase, glucoamylase and glucose isomerase are 500 s^{-1}, 160 s^{-1} and 3 s^{-1} respectively; an enzyme with a relative molecular mass of 60000 and specific activity 1 U mg^{-1} has a turnover number of 1 s^{-1}). The ratio k_{cat}/K_m determines the relative rate of reaction at low substrate concentrations, and is known as the specificity constant. It is also the apparent 2nd order rate constant at low substrate concentrations where,

$$v = \frac{K_{cat}}{K_m}[E]_0[S]$$

Many applications of enzymes involve open systems, where the substrate concentration remains constant, due to replenishment, throughout the reaction time-course. This is, of course, the situation that often prevails *in vivo*. Under these circumstances, the Michaelis-Menten equation is obeyed over an even wider range of enzyme concentrations than allowed in closed systems, and is commonly used to model immobilised enzyme kinetic systems.

Enzymes have evolved by maximising k_{cat}/K_m (i.e., the specificity constant for the substrate) while keeping K_m approximately identical to the naturally encountered substrate concentration. This allows the enzyme to operate efficiently and yet exercise some control over the rate of reaction.

The specificity constant is limited by the rate at which the reactants encounter one another under the influence of diffusion. For a single-substrate reaction the rate of encounter between the substrate and enzyme is about 10^8 - 10^9 M^{-1} s^{-1}. The specificity constant of some enzymes approach this value although the range of determined values is very broad (e.g., k_{cat}/K_m for catalase is 4 x 10^7 M^{-1} s^{-1}, whereas it is 25 M^{-1} s^{-1} for glucose isomerase, and for other enzymes varies from less than 1 M^{-1} s^{-1} to greater than 10^8 M^{-1} s^{-1}).

Enzyme Inhibition

Inhibition of specific enzymes by drugs can be medically useful. Understanding the mechanisms of enzyme inhibition is therefore of considerable importance. We will discuss four types of enzyme inhibition – competitive, non- competitive, uncompetitive, and suicide. Of these, the first three types are reversible. The last one is not.

Competitive Inhibition

Probably the easiest type of enzyme inhibition to understand is competitive inhibition and it is the one most commonly exploited pharmaceutically. Molecules that are competitive inhibitors of enzymes resemble one of the normal substrates of an enzyme. An example is methotrexate, which resembles the folate substrate of the enzyme dihydrofolate reductase (DHFR). This enzyme normally catalyzes the reduction of folate, an important reaction in the metabolism of nucleotides. When the drug methotrexate is present, some of the enzyme binds to it instead of to folate and during the time methotrexate is bound, the enzyme is inactive and unable to bind folate. Thus, the enzyme is inhibited. Notably, the binding site on DHFR for methotrexate is the active site, the same place that folate would normally bind. As a result, methotrexate 'competes' with folate for binding to the enzyme. The more methotrexate there is, the more effectively it competes with folate for the enzyme's active site. Conversely, the more folate there is, the less of an effect methotrexate has on the enzyme because folate outcompetes it.

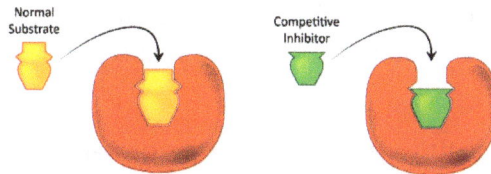

Figure: Competitive Inhibition

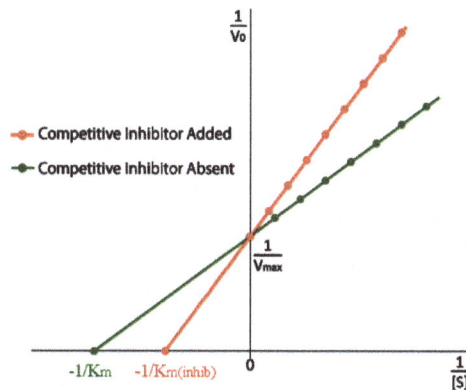

Figure: Line-Weaver Burk Plot of competitive inhibition

No Effect on V~MAX~

By performing a set of V vs. [S] reactions without inhibitor. competitive inhibition can be studieds V vs. [S] is plotted, as well as 1/V vs. 1/[S], if desired. Next, a second set of reactions is performed in the same manner as before, except that a fixed amount of the methotrexate inhibitor is added to each tube. At low concentrations of substrate, the inhibitor competes for the enzyme effectively, but at high concentrations of substrate, the inhibitor will have a much reduced effect, since the substrate outcompetes it, due to its higher concentration (remember that the inhibitor is at fixed concentration). at high substrate concentrations, the competitive inhibitor has essentially no effect, causing the Vmax for the enzyme to remain unchanged. To reiterate, this is due to the fact that at high substrate concentrations, the inhibitor doesn't compete well. However, at lower substrate concentrations it does.

Increased KM

the apparent KM of the enzyme for the substrate increases (-1/KM gets closer to zero - Red line in Line-Weaver Burk Plot of competitive inhibition) when the inhibitor is present, thus illustrating the better competition of the inhibitor at lower substrate concentrations. It may not be obvious why we call the changed KM the apparent KM of the enzyme. The reason is that the inhibitor doesn't actually change the enzyme's affinity for the folate substrate. It only appears to do so. This is because of the way that competitive inhibition works. When the competitive inhibitor binds the enzyme, it is effectively 'taken out of action.' Inactive enzymes have No affinity for substrate and no activity either. KM for an inactive enzyme cannot be measured.

The enzyme molecules that are not bound by methotrexate can, in fact, bind folate and are active. Methotrexate has no effect on them and their KM values are unchanged. Why then, does KM appear higher in the presence of a competitive inhibitor. The reason is that the competitive inhibitor is reducing the amount of active enzyme at lower concentrations of substrate. When the amount of enzyme is reduced, one must have more substrate to supply the reduced amount of enzyme sufficiently to get to $V_{max}/2$.

It is worth noting that in competitive inhibition, the percentage of inactive enzyme changes drastically over the range of [S] values used. To start, at low [S] values, the greatest percentage of the enzyme is inhibited. At high [S], no significant percentage of enzyme is inhibited. This is not always the case.

Non-Competitive Inhibition

A second type of inhibition employs inhibitors that do not resemble the substrate and bind not to the active site, but rather to a separate site on the enzyme. The effect of binding a non-competitive inhibitor is significantly different from binding a competitive inhibitor because there is no competition. In the case of competitive inhibition, the effect of the inhibitor could be reduced and eventually overwhelmed with increasing amounts of substrate. This was because increasing substrate made increasing percentages of the enzyme active. With non-competitive inhibition, increasing the amount of substrate has no effect on the percentage of enzyme that is active. Indeed, in non-competitive inhibition, the percentage of enzyme inhibited remains the same through all ranges of [S].

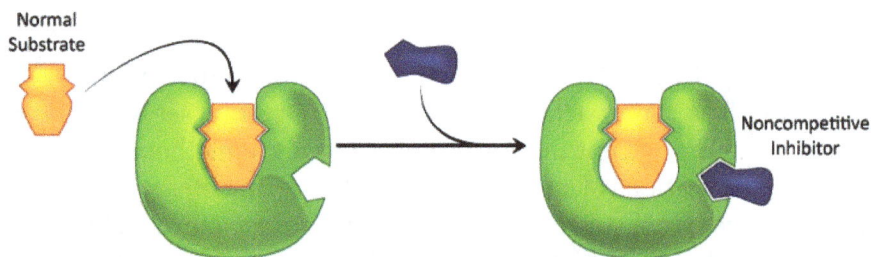

Figure: Non-competitive Inhibition

This means, then, that non-competitive inhibition effectively reduces the amount of enzyme by the same fixed amount in a typical experiment at every substrate concentration used The effect of this inhibition is shown above. V_{max} is reduced in non-competitive inhibition compared to uninhibited reactions. This makes sense if we remember that Vmax is dependent on the amount of enzyme present. Reducing the amount of enzyme present reduces Vmax. In competitive inhibition, this doesn't occur detectably, because at high substrate concentrations, there is essentially 100% of the enzyme active and the Vmax appears not to change. Additionally, KM for non-competitively inhibited reactions does not change from that of uninhibited reactions. This is because, one can only measure the KM of active enzymes and KM is a constant for a given enzyme.

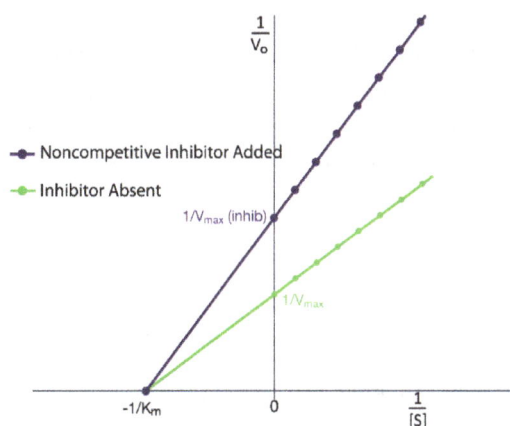

Figure: Line-Weaver Burk Plot of noncompetitive inhibition

Uncompetitive Inhibition

A third type of enzymatic inhibition is that of uncompetitive inhibition, which has the odd property of a reduced Vmax as well as a reduced KM. The explanation for these seemingly odd results is rooted in the fact that the uncompetitive inhibitor binds only to the enzyme-substrate (ES) complex. The inhibitor-bound complex forms mostly under concentrations of high substrate and the ES-I complex cannot release product while the inhibitor is bound, thus explaining the reduced V_{max}.

The reduced KM is a bit harder to conceptualize. The answer lies in the fact that the inhibitor-bound complex effectively reduces the concentration of the ES complex. By Le Chatelier's Principle, a shift occurs to form additional ES complex, resulting in less free enzyme and more enzyme in the forms ES and ESI (ES with inhibitor). Decreases in free enzyme correspond to an enzyme with greater affinity for its substrate. Thus, paradoxically, uncompetitive inhibition both decreases Vmax and increases an enzyme's affinity for its substrate.

Suicide Inhibition

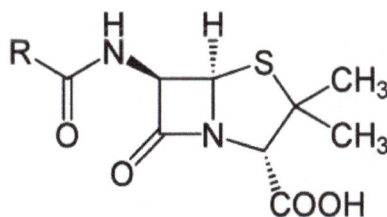

Figure: Penicillin

In contrast to the first three types of inhibition, which involve reversible binding of the inhibitor to the enzyme, suicide inhibition is irreversible because the inhibitor becomes covalently bound to the enzyme during the inhibition and thus cannot be removed. Suicide inhibition rather closely resembles competitive inhibition because the inhibitor generally resembles the substrate and binds to the active site of the enzyme. The primary difference is that the suicide inhibitor is chemically reactive in the active site and makes a bond with it that precludes its removal. Such a mechanism is that employed by penicillin, which covalently links to the bacterial enzyme, D-D transpeptidase and stops it from functioning. Since the normal function of the enzyme is to make a bond necessary for the peptido-glycan complex of the bacterial cell wall, the cell wall cannot properly form and bacteria cannot reproduce. If one were to measure the kinetics of suicide inhibitors under conditions where there was more enzyme than inhibitor, they would resemble non-competitive inhibition's kinetics because both involve reducing the amount of active enzyme by a fixed amount in a set of reactions.

References

- Intro, pages, mintermm, Webpapers: chem.uwec.edu, Retrieved 19-February-2019

- Glycoside_hydrolases: cazypedia.org, Retrieved March 17, 2019

- Lyase-Introduction, resource: creative-enzymes.com, Retrieved August 2, 2019

- Ligase, science: britannica.com, Retrieved March 4, 2019

- Enzymes, biomolecules, biology, guides: toppr.com, Retrieved April 21, 2019

- kinetics, enztech, water: lsbu.ac.uk, Retrieved January 7, 2019

- Enzyme_Inhibition, Catalysis, Biochemistry_Free_and_Easy, Biochemistry: libretexts.org , Retrieved March 14, 2019

Permissions

Index

A

Acyl Carrier Protein, 76, 78-80, 82-83
Adenosine Triphosphate, 62, 144, 147, 172-173, 178, 216
Adipose Cells, 26, 60-61, 113
Aldohexose, 11, 17, 20, 23
Amino Acids, 2-5, 7, 22, 24, 58, 65, 87-88, 91, 121-129, 132, 135-140, 142-143, 146-147, 150-151, 153-154, 156-162, 164-165, 168-169, 178, 186, 190, 192, 200, 210
Amphipathic Lipids, 60-61, 89
Anaerobic Respiration, 57-58
Arachidonic Acid, 68, 71, 74-75, 88, 117
Ascorbic Acid, 2, 18, 25

B

Bile Pigments, 139, 143

C

Carbohydrates, 2, 7-12, 14, 17-18, 22-29, 31, 37, 41-42, 47, 54-55, 60, 63-64, 78-79, 95, 121-122, 147, 150, 152, 159, 166, 168, 170, 191, 200
Catalytic Triad, 92, 119
Cellulose, 5, 10-12, 14, 16, 23, 25, 38-42, 45, 50, 55, 131, 137
Cholesterol Biosynthesis, 98-103
Cholesterol Esters, 85-88, 91-92, 96, 102, 104-107, 112, 115-117
Citric Acid Cycle, 25, 28, 34, 56-57, 64, 168, 215
Coenzyme A, 62, 76, 95, 107-108, 110-111, 119, 207
Coprastanol, 102, 107
Cytochrome, 4, 136
Cytosolic Enzyme, 118
Cytosolic Lipid, 104, 114

D

Deoxyribonucleic Acid, 2, 18, 67, 150, 154, 171-172, 181
Dihydroxyacetone Phosphate, 28, 56-57, 93, 107, 215
Disaccharide Sucrose, 10, 26, 33
Disaccharides, 9-10, 16, 22, 31-33, 37-38, 210
Disulfide Bond, 87-88, 90, 124, 130, 141, 149, 157

E

Eicosanoids, 67-69, 72-75, 107, 113, 115, 117
Endoplasmic Reticulum, 72, 76, 83, 90, 92, 96, 99-100, 102-105, 108-111, 113-115, 118-119, 164
Eukaryotes, 67, 106, 109, 174, 176-178, 183-184, 187-188

F

Fatty Acid Oxidation, 64, 76-78, 83-84
Flavin Adenine Dinucleotide, 100
Free Fatty Acids, 61-62, 64, 69, 71, 76, 83, 85, 91-92, 106-107, 110-112, 115-116, 119
Fructose, 9-10, 13-14, 16-18, 24, 26-28, 31-34, 37-38, 47-48, 54-55, 57, 59, 159, 203, 205-206, 209, 215

G

Galactose, 9-10, 16, 18, 24, 26, 28-32, 34-36, 38, 152, 210, 212-213
Galactosemia, 29, 31, 212-213
Gastric Juice, 3, 150, 152
Gene Expression, 67, 72, 83, 91, 187-188
Glucose Metabolism, 11, 53
Glycosidic Bonds, 32, 37, 40, 43, 47, 51
Glycosidic Linkages, 43, 51, 53

H

Hexokinase, 28, 54-56, 59
High-density Lipoproteins, 85, 102, 104, 113, 152
Hydrochloric Acid, 3, 128, 135, 158, 216
Hydroxide Ion, 128
Hypoglycemia, 26-27

K

Ketohexose, 11, 17-18, 27
Ketone Group, 23
Krebs Cycle, 34, 54, 56-58, 63, 79, 95, 215

L

Lactose, 10, 12, 22-23, 29-32, 34-38, 150, 210, 213
Leloir Pathway, 29-31
Lipid Core, 89, 114
Lipid Metabolism, 60, 89, 93-95, 97, 106, 109, 112-113, 116-117
Liver X Receptor, 94, 103
Low-density Lipoproteins, 74, 76, 85, 95, 101, 113
Lysosomes, 5, 91, 164, 167

M

Maltodextrin, 47
Mitochondria, 5, 34, 56-58, 62, 65, 70, 72, 76, 78-79, 95-96, 109, 111, 114, 164, 168, 212
Mitochondrial Dna, 178-179

Monosaccharides, 3, 7, 9-11, 13, 18, 21-22, 29, 31-33, 38-40, 55, 210

Myristic Acid, 70, 78

N

Nucleic Acids, 1-2, 4, 6, 8-9, 25, 60, 122, 143, 171, 177, 185

O

Oleic Acid, 70, 74, 111

Oligosaccharides, 9-10, 16, 152, 169

One Glucose Molecule, 13, 16, 56-57

Oxidative Phosphorylation, 34, 54, 63-64, 178, 191-192

Oxidizing Agent, 46, 81

P

Palmitic Acid, 63, 70, 78, 112

Pancreatic Lipase, 92, 110-112

Pentose Phosphate Pathway, 25, 51, 54, 81, 171, 209

Peptide Bonds, 123, 126, 128, 132, 135, 139, 141-142, 144, 148, 162, 209

Plasma Lipid, 27, 93, 106

Polysaccharides, 3, 7, 9-14, 18, 22, 25, 31, 39-41, 44, 55, 60, 210

Polyunsaturated Fatty Acids, 68-71, 78

Protein Kinase C, 68, 95

Proteolipids, 85, 97, 152

R

Resistant Starch, 45, 47, 49

Ribonucleic Acid, 2, 154-155, 171-172

S

Stereoisomerism, 14, 17

Sterol, 60, 83, 94, 97-98, 100, 102-103, 106

Sucrose, 10, 12, 23, 26, 28-29, 31-34, 37-38, 48, 206

T

Thromboxanes, 68, 73

Tricarboxylic Acid Cycle, 56-57, 77, 79

Triglycerides, 8, 60-62, 95-96

X

X-ray Crystallography, 6, 35, 46, 197, 207

X-ray Diffraction, 133, 137-139